Logical Fallacies of Special and General Theory of Relativity

Second Edition:

General Relativity Doesn't Meet the Principles of Scientific Method

GATOT SOEDARTO

Copyright © 2017 Gatot Soedarto

All rights reserved.

Including the right to reproduce this book or portions thereof, in any form. No part of this text may be reproduced in any form without the express written permission of the author

ISBN-13: **978-1544735313**
ISBN-10: **1544735316**

Printed in USA by CreateSpace Independently Book Publisher
Last update: August 28, 2017

4.950 KB

DEDICATION

I dedicate this book for Nanda-Caca-Radeva, Salsa-Sherly, Satriya-Ayu, and Abel Abel-Afnan..

CONTENTS

Acknowledgments

Preface 6

1: Logical Fallacies of Einstein's Theories. 9

1.1. Thought Experiment, Spacetime,

and EFE. 9

1.2. Testing Einstein's Hypothesis. 38

1.3. Astronomical Test of General Relativity 66

1.4. Astronomical Data Proves Spacetime is False. ... 84

2. Experiments: Knowing the Result They Wanted to Get 90

3. What Exactly is Gravity? 153

4. Conclusion 162

Reference

About the Author

ACKNOWLEDGMENTS

I would like to express my gratitude to the many people who saw me through my books; to all those who provided support, read, buy, and offered comments.

My claim:
Einstein's hypothesis of general relativity is not valid; and Einstein proposed test via eclipse is not scientifically correct and deeply wrong. He wants measuring deflection of light by the Sun but he proposed test to measure deflection of light by the Earth's atmosphere. General relativity can not be proven or tested in any way.

PREFACE

Einstein presented Special Theory of Relativity in the year 1905 to explain that the speed of light must be the same for all observers, regardless of their velocities. Einstein began by rejecting the ether theory without any proof whatsoever; he took the idea from the Michelson-Morley experiment. But in 1920 Einstein has accepted the ether theory. Albert Einstein gave an address on 5 May 1920 at the University of Leiden: "Recapitulating, we may say that according to the general theory of relativity space is endowed with physical qualities; in this sense, therefore, there exists ether. According to the general theory of relativity space without ether is unthinkable; for in such space there not only would be no propagation of light, but also no possibility of existence for standards of space and time ..." Special relativity rejects the ether while general relativity requires the ether. That is something incompatibility between special and general relativity.

This book is 2nd edition of previous book entitle Logical Fallacies of Special and General Theory of Relativity. I've found in the Lincoln Barnett's book 'The Universe and Dr. Einstein', London, 1949, foreword by Albert Einstein himself; the back ground of Einstein's spacetime idea. It seems that Einstein had no idea on the basic of astronomy; and not knowing that space and time has been applied in astronomy- as the celestial sphere coordinates system-for long time ago. Einstein admired to Riemann geometry and 4 D (dimension)

Minkovsky space, without considering the fact that Riemann geometry and Minkovsky space were not applied in astronomy. Beside that, I've found at least 5 logical fallacies of Einstein's theory of relativity.

The most important thing, I've found Einstein's hypothesis of general relativity is not valid. Ironically, Einstein proposed test via eclipse is not scientifically correct, and deeply wrong. Hypothesis and Einstein proposed test of general relativity are closely related to astronomy, especially celestial navigation. For understanding that hypothesis and the test are not valid, physics training is needed; but more importantly is celestial navigation training. Unfortunately, physicists and astrophysicists are not trained to become experts in the field of celestial navigation. The navigators around the world will be easily to recognize the fatal flaws of these hypotheses and test. Actually, general relativity can not be proven or tested in any way.

No one will tell you about this, but I will....

Stephen Hawking said: "Some people never admit that they are wrong and continue to find new, and often mutually inconsistent, arguments to support their case ". I will tell you that all tests which says 'general relativity is correct' really are the case of 'knowing the result they wanted to get'

I believe to scientist who said: "If you disagree because of who said it, not because of what was said, you're part of the problem."

A physics professor said that given Einstein's status as a popular icon, there are countless people who wish to prove him wrong, even among scientists with degrees to their names. Does it mean that one can not reveal Einstein's fault, although the evidence and fact had been found that his theory is invalid?

I think, it doesn't matter people wish to prove Einstein was wrong with the goal to their reputation or not, because many people will test the findings. If the findings are incorrect, it will further enhance Einstein's status as a popular icon. If the findings are correct, it has very important for the future generations of scientists.

August 28, 2017

Capt (Ret) Gatot Soedarto

Navigator, former lecturer on Astronomy.

1. LOGICAL FALLACIES OF EINSTEIN'S THEORIES

1.1. Thought Experiment, Spacetime, and EFE.

Einstein's elevator

Albert Einstein used a thought experiment to describe his unique approach of the basic concepts of his ideas on theory of gravity. One of his thought experiment is about elevator.

Figure 1.1: According to general relativity, objects in a gravitational field behave similarly to objects within an accelerating enclosure. For example, an observer will see a ball fall the same way in a rocket (left) as it does on Earth (right), provided that the acceleration of the rocket is equal to 9.8 m/s2 (the acceleration due to gravity at the surface of the Earth).-(en.wikipedia.org)

Lincoln Barnett's book, The Universe and Dr.Einstein, page 69, makes familiar to us:

"This physicists are still in the elevator, but this time they really are in the empty space, far away from the attractive power of any celestial body. A cable is attached to the roof of the elevator; some supernatural force begins reeling in the cable; and the elevator travels "upward" with constant acceleration, i.e. progressively faster and faster. Again the men in the car have no idea where they are, and again they perform experiments to evaluate their situation. This time they notice that their feet press solidly against the floor come up beneath them.

If they release objects from their hands, the objects appear to "fall". If they toss object in a horizontal direction they do not move uniformly in a straight line, but describe a parabolic curve with respect to the floor.

> Einstein now shifts the scene. The physicists are still in the elevator, but this time they really are in empty space, far away from the attractive power of any celestial body. A cable is attached to the roof of the elevator; some supernatural force begins reeling in the cable; and the elevator travels "upward" with constant acceleration, i.e. progressively faster and faster. Again the men in the car have no idea where they are, and again they perform experiments to evaluate their situation. This time they notice that their feet press solidly against the floor. If they jump they do not float to the ceiling, for the floor comes up beneath them. If they release objects from their hands, the objects appear to "fall." If they toss objects in a horizontal direction they do not move uniformly in a straight line, but describe a parabolic curve with respect to the floor. And so the scientists, who have no idea that their windowless car actually is climbing through interstellar space, conclude that they are situated in quite ordinary circumstances in a stationary room rigidly attached to the earth and affected in normal measure by the force of gravity. There is really no way for them to tell whether they are at rest in a gravitational field or ascending with constant acceleration through outer space where there is no gravity at all.

Figure 1.2: The Universe and Dr. Eintein, London 1949, page 69.

And so the scientist, who have no idea that their windowless car actually is climbing through interstellar space, conclude that they are situated in quite ordinary circumstances in a stationary room rigidly attached to the earth and affected in normal measure by the force of gravity. There is no way for them to tell whether they are at rest in a gravitational field or ascending with constant acceleration through outer space where there is no gravity at all."

> So Einstein's Law of Gravitation contains nothing about force. It describes the behaviour of objects in a gravitational field—the planets, for example—not in terms of "attraction" but simply in terms of the paths they follow. To Einstein, gravitation is simply part of inertia; the movements of the stars and the planets arise from their inherent inertia; and the courses they follow are determined by the metric properties of space—or, more properly speaking, the metric properties of the space-time continuum.

Figure 1.3: The Universe and Dr.Eintein, London 1949, page 70

"So Einstein's Law of Gravitation contain nothing about force. It describes the behavior of objects in a gravitational field—the planets, for example—not in terms "attraction" but simply in the terms of the paths they follow. To Einstein, gravitation is simply part of inertia; the movements of stars and the planets arise from their inherent inertia; and the courses they follow are determined by the metric properties of space—or, more properly speaking, the metric properties of the space-time continum." (The Universe and Dr.Eintein, Lincoln Barnett, London 1949, page 69–72). [1]

Einstein's elevator gives an illustration about the idea of the effects of gravity:

1. They notice that their feet press solidly against the floor come up beneath them.

2. If they release objects from their hands, the objects appear to "fall".

3.. If they toss object in a horizontal direction they do not move uniformly in a straight line, but describe a parabolic curve with respect to the floor.

From the above summary we get the clear picture that the three points of ideas in the elevator are about the effects of gravity to the objects appear to fall. In the Einstein's elevator nothing about the effect of gravity to the planets.

Fallacy of composition

It clear that the effect found must be attributed to objects inside the elevator, and not, to the planets-objects outside the elevator. But an observer outside the elevator drew a conclusion that gravity is nothing about force, and the behavior of objects in a gravitational field—the planets, for example—not in terms "attraction" but simply in the terms of the paths they follow.

: An observer outside the elevator had not realized that the thought experiment is incomplete. Three objects of observations inside the elevator can not describe the complete meaning of gravity, for example, can not describe at least two effects that are caused by the force of gravity

1. The celestial bodies in orbit, for example, the orbiting of the planets around the Sun.

2. The occurrence of tide-low tide caused by the attraction of the moon.

Of course it will be very difficult to illustrate two things mentioned above in the elevator or in single one of thought experiment, not one by one of thought experiment.

What could be illustrated in the elevator is only related to the objects that have weight and mass.

That is fallacy of composition, assuming that something true of part of a whole must also be true of the whole A fallacy is an incorrect argument in logic and rhetoric which undermines an argument's logical validity or more generally an argument logical soundness. Therefore, the Einstein's thought experiment are incomprehensive, illogical, and misleading.

Actually, thought experiments can be made to obtain the result they wanted to get. In this case we see that Albert Einstein made a mistake in his thought experiments of elevator.

The Equivalence Principle

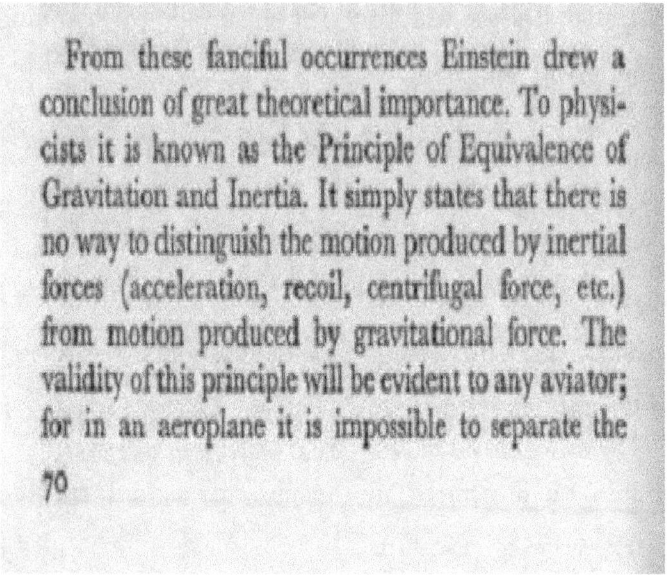

> From these fanciful occurrences Einstein drew a conclusion of great theoretical importance. To physicists it is known as the Principle of Equivalence of Gravitation and Inertia. It simply states that there is no way to distinguish the motion produced by inertial forces (acceleration, recoil, centrifugal force, etc.) from motion produced by gravitational force. The validity of this principle will be evident to any aviator; for in an aeroplane it is impossible to separate the
>
> 70

Figure 1.4: The Universe and Dr.Eintein, London 1949, page 70.

"From these fanciful occurrences Einstein drew a conclusion of great theoretical importane. To physics it is known as the principles of equivalence of gravitation and inertia. It is simply states that there is no way to distinguish the motion produced by inertial forces (acceleration, recoil, centrifugal force, etc) from motion produced by gravitational force. *The validity of this principle will be evident to any aviator;* ..."

The validity of this principle will be evident to any aviator, of course not, the problem here is not merely that the objects appear to fall

Thought experiment isn't the real experiment. All of the thought experiments can be made to obtain the result they wanted to get, and can be misleading. Moreover, from the thought experiment that are incomprehensive and incorrect argument in logic Einstein drew a conclusion of the equivalence principle. This principle does not meet the requirements of scientific theory that are should be based on the real experiment.

That is a logical fallacy, a false equivalence – describing a situation of logical and apparent equivalence, when in fact there is none That is why the equivalence principle is false.

Bending of light

. Einstein's idea that light travel as a curve or bending of light in the gravity field of massive object, in accordance with his equivalence principle and on another thought experiment. See Figure 1.5.

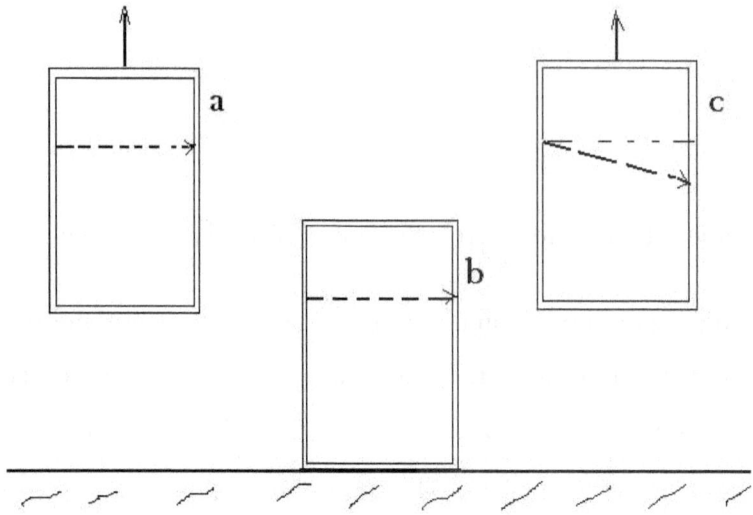

Figure 1.5: Bending of light

In Figure (a), the elevator is accelerating upwards, and according to equivalence principle this elevator is the same as in Figure 1(b) elevator on the Earth.. Viewed from outside, a laser beam follows a straight line in accordance with Newton's theory.

But in Figure (c), viewed inside the elevator, the light beam appears to curve downwards. The effect in a stationary elevator situated in a gravitational field is the same, as in Figure (b).

An observer inside the elevator drew a conclusion that light travel as a curve or bending of light, it's nonconformity with Newton's theory.

Actually, that is something logical fallacy, a false equivalence –

describing a situation of logical and apparent equivalence, when in fact there is none. Remember that a thought experiment can be made to obtain the results they wanted to get. In the above thought experiment they wanted to get the bending of light.

We can use the above experiment to explain that a laser beam follows a straight line in accordance with Newton's theory. See Figure 1.6.

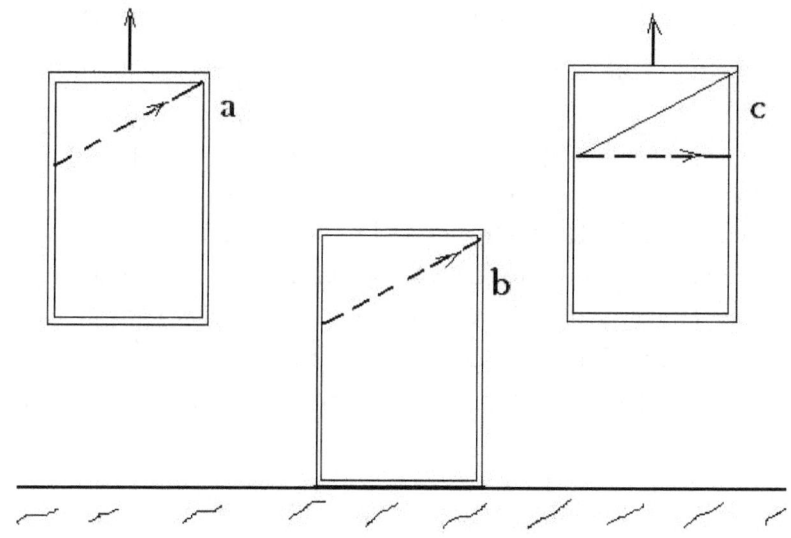

Figure 1.6: No bending of light

In Figure (a), the elevator is accelerating upwards, and according to equivalence principle this elevator is the same as in Figure 1(b) elevator on the Earth.. Viewed from outside, a laser beam follows a straight line in accordance with Newton's theory.

And in Figure (c), viewed inside the elevator, the light beam remain appears in a straight line, no bending of light. The effect in a stationary elevator situated in a gravitational field is the same, as in Figure (b).

An observer inside the elevator drew a conclusion that light travel as a straight line in accordance with Newton's theory.

From the thought experiment Einstein also drew a conclusion of the metric properties of the space-time continuum, or spacetime. Spacetime is the basic of Einstein's theory about gravity.

So, it is clear that spacetime does not meet the requirements of scientific theory, because it's come from thought experiment, not the real experiment. In facts, a scientist looking at spacetime is the same with atmospheric medium, for example in the case of satellite Gravity Probe B. The goal of NASA's Gravity Probe B to prove general relativity, .and this satellite has an orbit at altitude 642 km above the Earth. Its mean this orbit is in the Earth's atmosphere. This will be further discussed later.

What is a spacetime?

According to Einstein, as special theory of relativity showed, space and time separately are relative quantities which vary with individual observers. Einstein take this idea from Minkovski, who developed the mathematics of the space-time continuum. Minkovski's spacetime (4 D) is a mathematical construction.

But for Einstein, the space-time continuum is not simply a mathematical construction. Einstein viewed the world is a space-time continuum.

"To describe any physical event involving motion, however, it is not enough simply to indicate position in space. It is necessary to state also how position changes in time. Thus to give an accurate picture of the operation of New York - Chicago express, one must mention not only that is goes from New York to Albany to Siracuse to Cleveland to Toledo to Chicago, but also the times at which it touches each of those points. his can be done either by means of a timetable or a visual chart.

In same way the flight of an aeroplane from New York to Los Angeles can best be pictured in a four-dimensional space-time continuum. The fact that the plane is at latitude x, longitude y, and altitude z, means nothing to the traffic manager of the airplane unless the time co-ordinate is also given. So time is fourth dimensional. And if one wishes to envisage the flight as a whole, as a physical reality, it can not be broken down into a series of disconnected take-offs, climbs, glides, and landing. Instead it must be thought of as continuum curve in a four-dimensional space-time continuum.

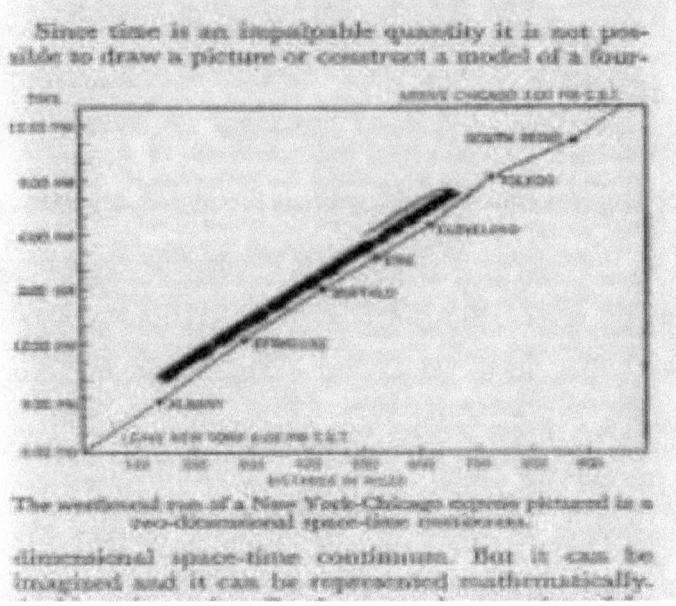

Figure 2.1.

Since time is an impalpable quantity it is no possible to draw a picture or construct a model of a four-dimensional space-time continuum. But it can be imagined and it can be represented mathematically.

Obviously the astronomers has to think of the Universe as space-time continuum" (Lincoln Barnett, The Universe and Dr.Einstein, page 58-61)

Einstein advises to the astronomers of the need to take into account the time, because time can not be separated by space. Let us note : Einstein advises to the astronomers about the need to take into account the time. Here is a question: Why didn't he know space and time are combined together in astronomy almost forever?

About spacetime, Stephen Hawking also explained in the background of the Space and Time Warps.

"General Relativity was a major intellectual revolution that has transformed the way we think about the universe. It is a theory not only of curved space, but of curved or warped time as well.

Einstein had realized in 1905, that space and time, are intimately connected with each other. One can describe the location of an event by four numbers. Three numbers describe the position of the event. They could be miles north and east of Oxford circus, and height above sea level. On a larger scale, they could be galactic latitude and longitude, and distance from the center of the galaxy. The fourth number is the time of the event. Thus one can think of space and time together, as a four-dimensional entity, called space-time. Each point of space-time is labeled by four numbers that specify its position in space, and in time. Combining space and time into space-time in this way would be rather trivial, if one could disentangle them in a unique way.

That is to say, if there was a unique way of defining the time and position of each event. However, in a remarkable paper written in 1905, when he was a clerk in the Swiss patent office, Einstein showed that the time and position at which one thought an event occurred, depended on how one was moving. This meant that time and space, were inextricably bound up with each other."(Stephen Hawking, hawking.org.uk).

"Thus one can think of space and time together, as a four-dimensional entity, called space-time."

Excerpt and the above statement explain the background of Einstein's ideas about spacetime. Obviously, the ideas arise from misunderstanding or rather ignorance of basic astronomy. Einstein had not realized that astronomy use non-Euclidean geometry: spherical geometry. It is the same with Einstein's ideas on general relativity use non-Euclidean geometry. But Einstein's non-Euclidean geometry is not spherical geometry.

Spherical /non-Euclidean geometry is 3 D + 1 D. In this term, 3 D is space dimension, and 1 D is time dimension or to be more precisely 'time direction'. The Universe is three-dimension that has time dimension as directions - neither up nor down, left nor right, in nor out.

Einstein made a new term 4D, which is known as spacetime. But 4D spacetime has no direction. A dimension system must have a direction. Supposedly 4 D spacetime and 1 dimension / direction - 5th dimensions - but this is impossible.

From the discussion above, one can conclude that Einstein had no idea on the basic of astronomy. This will be more clearly in the next discussion.

In astronomy space and time are combined together, but not as a four-dimensional entity. Space and time are combined almost forever, which is applied in the Equatorial Coordinate System, in the form of Hour Circle, Hour Angle, Local Hour Angle (LHA), Greenwich Hour Angle (GHA), and Right Ascension (RA). The Equatorial Coordinate System is based on The Horizontal Coordinate System. In this coordinate system, we can determine the position of a certain star by knowing the azimuth and altitude of the stars.

Azimuth and Altitude of stars change at any time due to the daily movements of the stars, and writing azimuth and altitude are in degree, minute, and seconds.

Suppose, then, that you measure the angle of the Star above the horizon -the altitude of Star, and find it is x degrees. In celestial sphere coordinates system, x degree equals x hour and x hour equals 60 nautical mile. So, in astronomy we can transform .degree, minute, seconds of an arc in time: hours, minute, seconds and we can also transform degree in distance to miles and yards.

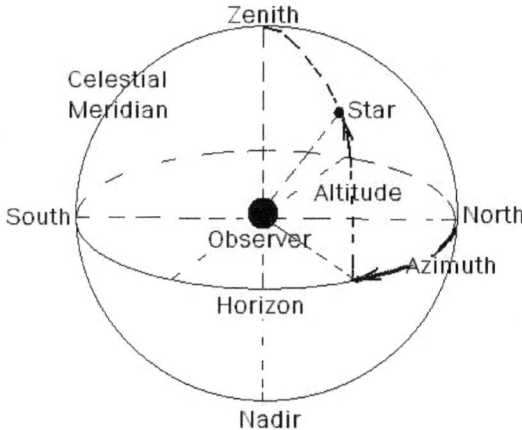

THE HORIZONTAL COORDINATE SYSTEM

Figure 2.2: The Horizontal Coordinate Syestem

In astronomy, there are 4 celestial sphere coordinates system: The Horizontal System, Equatorial system, Ecliptic system, and Galactic system.

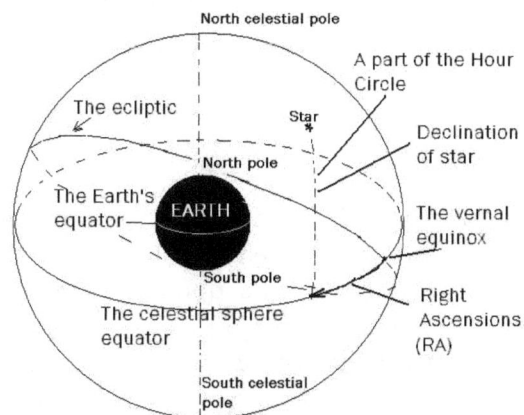

THE EQUATORIAL COORDINATE SYSTEM

Figure 2.3: The equatorial coordinate system

The Equatorial Coordinate System

The equatorial coordinate system is formed by projection of the Earth's equator onto the celestial sphere, forming the celestial equator.

Declination.

The declination is the angular distance of star to the celestial equator, positive to the North, negative to the South. Declination is measured in degrees (°), arc-minutes (') and arc-seconds ("), and is similar to latitude on Earth. Declination is analogous to Latitude. The declination of star is a part of the Hour Circle.

Right Ascention (RA):

The Right Ascension is the angular distance of star Eastward along the celestial equator from the vernal equinox to the Hour Circle passing through the star. Right Ascension is measured in hours, minutes, and seconds, and is similar to longitude on Earth

The Hour Circle is the great circle through the star and the Celestial Poles.

In this case, it seems clear that Einstein had no idea of the celestial sphere coordinates system. If he understand the fundamental concept of astronomy, of course, he don't want to give advice to the astronomers of the need to take into account the time, because time can not be separated by space. That is something logical fallacies of Einstein ideas. Actually, the celestial sphere coordinates system is an imaginary or a model 2 D + 1 D space and time, it is not the same with Minkovski 's 4 D space-time'

Actualy, there's nothing complicated about the celestial sphere coordinates system. All it is a system that measures the angle of celestial bodies in the sky. Before Einstein life, astronomers have said that planets move in its particular destiny, and they had familiar with measuring the angle in degree, minute, and arc or in time: hour, minute, and seconds.

Again, the fallacy of Einstein

In physical reality, there is no such thing as a substance call spacetime. But, it seems Einstein confuses about the properties of the model with the properties of real space, as he wrote in 1916:

"Two celestial bodies in orbit will generate invisible ripples in spacetime" (Einstein manuscript of 1916).

Again, here is a question:" What is ripples in spacetime?" There is no such thing as spacetime in reality, but Einstein making it as real. Now, ripples in spacetime are known as gravitational waves. Einstein predicted gravitational waves in the year of 1916. But, actually the terminology 'ripples in spacetime' is a fallacy of ambiguity or reification.

Reification generally refers to making something real, bringing something into being, or making something concrete. Reification (also known as concretism, or the fallacy of misplaced concreteness) is a fallacy of ambiguity, when an abstraction (abstract belief or hypothetical construct) is treated as if it were a concrete, real event, or physical entity. In other words, it is the error of treating as a concrete thing, something which is not concrete, but merely an idea.

Another common manifestation is the confusion of a model with reality. Mathematical or simulation models may help understand a system or situation but real life will differ from the model (e.g. 'the map is not the territory. [2]

One of the most common reified models the general public may be familiar with looks something like this:

Figure 2.4: **A curvature of spacetime**

Over time, scientists have taken to assuming that bending spacetime is the actual cause of gravity, while forgetting that spacetime is merely a descriptor, not an actual causal agent that can do real work.

From the model of a curved space Einstein ignored the atmospheric medium around the massive bodies in space. In fact, the fabric of spacetime which curves around the mass is not the empty vacuum but the atmospheric medium. There are lot of physicists explain a curved geometry of spacetime, and they assuming that the massive bodies the Earth and planets just like a ball without atmospheric medium.

Atmospheric medium of the Earth are troposphere, stratosphere, mesosphere, thermosphere, and exosphere.

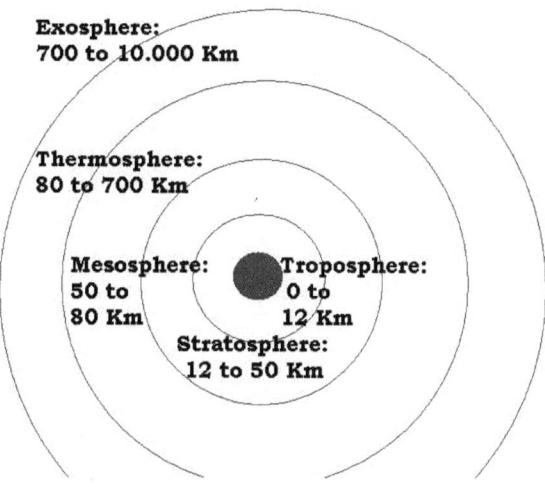

Figure 2.5: Earth's atmosphere

The stellar atmosphere is the outer region of the volume of a star, lying above the stellar core, radiation zone and convection zone. It is divided into several regions of distinct character: photosphere, chromospheres, and the outermost part of the stellar atmosphere is the corona, a tenuous plasma which has a temperature above one million Kelvin.

Mathematically Used by Einstein.

In 1905, Einstein proposed Special Relativity, and he began by rejecting the ether theory.

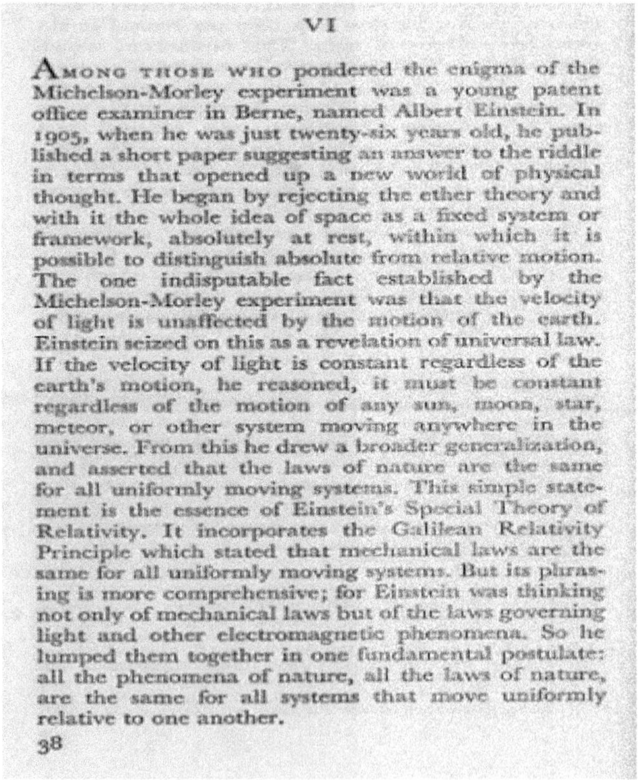

> **VI**
>
> Among those who pondered the enigma of the Michelson-Morley experiment was a young patent office examiner in Berne, named Albert Einstein. In 1905, when he was just twenty-six years old, he published a short paper suggesting an answer to the riddle in terms that opened up a new world of physical thought. He began by rejecting the ether theory and with it the whole idea of space as a fixed system or framework, absolutely at rest, within which it is possible to distinguish absolute from relative motion. The one indisputable fact established by the Michelson-Morley experiment was that the velocity of light is unaffected by the motion of the earth. Einstein seized on this as a revelation of universal law. If the velocity of light is constant regardless of the earth's motion, he reasoned, it must be constant regardless of the motion of any sun, moon, star, meteor, or other system moving anywhere in the universe. From this he drew a broader generalization, and asserted that the laws of nature are the same for all uniformly moving systems. This simple statement is the essence of Einstein's Special Theory of Relativity. It incorporates the Galilean Relativity Principle which stated that mechanical laws are the same for all uniformly moving systems. But its phrasing is more comprehensive; for Einstein was thinking not only of mechanical laws but of the laws governing light and other electromagnetic phenomena. So he lumped them together in one fundamental postulate: all the phenomena of nature, all the laws of nature, are the same for all systems that move uniformly relative to one another.
>
> 38

Figure 3.1: The Universe and Dr.Einstein, page 38

Lincoln Barnett's book, The Universe and Dr.Einstein, page 38, informs us:

"He began by rejecting the ether theory and with it the whole idea of space as a fixed system or framework, absolutely at rest, within which it is possible to distinguish absolute from relative motion. The one indisputable fact established by the Michelson-Morley experiment was that the velocity of light in unaffected by the motion of the earth."

Einstein rejects Ether Theory

The classical scientists such as Aristotle, Rene Descartes, Sir Isaac Newton and others believed that the light of the stars reaching us on earth crept spreading through a medium the so-called luminiferous ether. However various kinds of experiments had been made, among other was an experiment conducted by the American Scientists Michelson and Morrey 19th century, and all of those experiments failed to detect the presence of luminiferous ether, so that the ether is deemed non-existent.

There is a possibility that luminiferous ether truly exists, but it cannot be proven.

In 1905, on special theory of relativity, Einstein rejected Ether. But in 1920, in the general theory of relativity, Einstein accepted Ether. Albert Einstein gave an address on 5 May 1920 at the University of Leiden, and he said:

"Recapitulating, we may say that according to the general theory of relativity space is endowed with physical qualities; in this sense, therefore, there exists an ether. According to the general theory of relativity space without ether is unthinkable; for in such space there not only would be no propagation of light, but also no possibility of existence for standards of space and time ..."

In the views of logic, Einstein's statement is contradictory or contrary arguments. It's a type of logical fallacy. Contradictory argument, if one is correct, the other is not true. In the case of Einstein's theory, if general relativity is correct, special relativity can not be true. Or it can be a contrary argument, special and general relativity cannot both be true but can both be false.

"Einstein seized on this as a revelation of universal law. If the velocity of light is constant regardless of the earth's motion, he reasoned, it must be constant regardless of motion of any Sun, moon, star, meteor, or other syatem moving anywhere in universe.From this he drew a broader generalization, and asserted that the laws of nature are the same for alls uniformly moving system. This simple statement is the essence of Einstein's Special theory of Relativity."

A Revelation of Universal law

Einstein states that the velocity of light is constant. In this case, Einstein drew the conclusion from the Michelson-Morley experiment, but he had no idea of basic astronomy about the effect of light's refraction in the real world. It seems clear Einstein know not that the starlight or Sunlight can be slowed down by Earth's atmosphere, and gave an impact on time dilation of starlight.or Sunlight.

Figure 32 makes familiar to us about the effects of refraction of light: Let's note that A - B a little longer than A – C . It proves that time dilation of light is caused by refraction. In the other word, time dilation of light can be explained without Einstein's relativity.

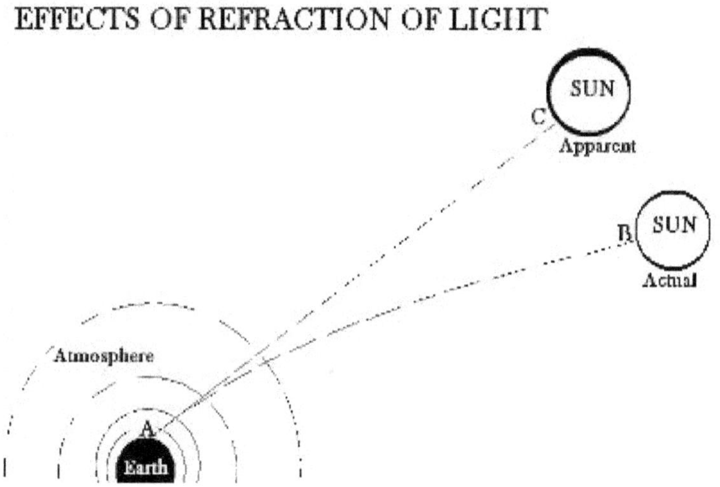

EFFECTS OF REFRACTION OF LIGHT

Figure 3.2: Time dilation of light.

Actually, Einstein's ideas about a revelation of Universal law: the velocity of light is constant and nothing can travel faster than the speed of light, both **are the logical fallacies.**

"From this he drew a broader generalization, and asserted that the laws of nature are the same for alls uniformly moving system", it is clear something a logical fallacy, a false equivalence – describing a situation of logical and apparent equivalence, when in fact there is none

"Nothing can travel faster than the speed of light."

"Light always travels at the same speed."

Have you heard these statements before? They are often quoted as results of Einstein's theory of relativity. *Unfortunately, these statements are somewhat misleading.* Let's add a few words to them to clarify. "Nothing can travel faster than the speed of light in a vacuum." "Light in a vacuum always travels at the same speed." Those additional three words in a vacuum are very important. A vacuum is a region with no matter in it. So a vacuum would not contain any dust particles. [3]

It can be said that Einstein started with no ideas on basic astronomy, this led Einstein unwittingly ignored refraction of light.and made a logical fallacy.

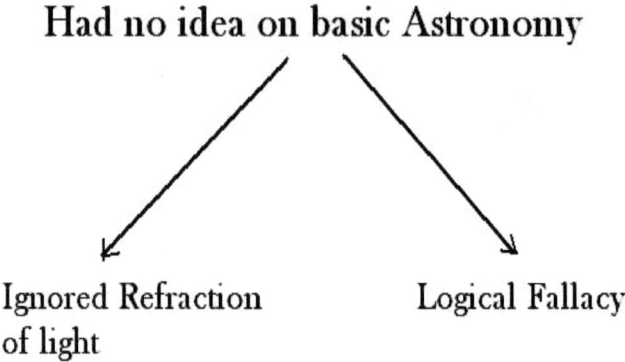

Figure 3.3

Einstein's Mathematics

The most famous Einstein's equation is $E = mc^2$, but we can found some of error in this equation.

"Einstein erred in his 'proof' of his most famous equation, $E = mc^2$. Einstein was only able to derive $E = mc^2$ for a particle completely at rest. Despite also inventing special relativity—founded on the principle that the laws of physics are independent of an observer's frame of reference—Einstein's formulation couldn't account for how energy worked for a particle in motion." (Ethan Siegel, The Four Biggest Mistakes Of Einstein's Sientific Life, Medium).

Einstein's special theory of relativity is made based on the equation developed by the great Dutch physicist H.A.Lorentz. Although its original application is of interest, now chiefly to scientific historian, the Lorentz transformation lives on as part of the mathematical frame-work of Einstein's relativity.

Unfortunately, mathematical procedure by which Albert Einstein derived Lorentz transformation is incorrect. The transformation is an imaginary "solution" to a set of equations which evaluate to zero throughout the derivation process.

For example error 1: Expression (3) is useless

$x - ct = 0$ (1)

$x' - ct' = 0$ (2)

In (3) Einstein writes

$x' - ct' = a(x - ct)$ (3) $a = $ lambda

Because of (1) and (2) we can write (3) as

$0 = a0$

One can postulate that meaningful values, which are at least sometimes both different from zero, are somehow related. Introducing proportionality quotient between nothing and nothing has no meaning. [4].

Einstein Field Equation of Gravitation (EFE).

Einstein's field equation (EFE) is usually written in the form:

$$R_{\mu\nu} - \frac{1}{2} R\, g_{\mu\nu} + \Lambda\, g_{\mu\nu} = \frac{8\pi G}{c^4} T_{\mu\nu}$$

The left-hand side of that equation is a matrix of numbers represents the curvature of spacetime as determined by the metric. The right-hand side is a matrix of quantum operators represents the matter/energy content of spacetime. Einstein's equations connect mass to spacetime curvature. The mathematical quantity associated with space-time curvature is the Riemann tensor.

In science, mathematics is used to describes the real world phenomena, to calculate, and to measure of quantity, distance, speed, force, weight, capacity … ..etc. Mathematics is widely used in physics, astronomy, chemical, and engineering, and helping us to understand about real world phenomena.

As the language of physics, there are at least four things logical fallacies in Einstein field equation of gravitation:

First, argument from silence – where the conclusion is based on the absence of evidence, rather than the existence of evidence. In this case, there is no evidence that spacetime is the real world. It is widely known that space vacuum has no properties.

Second, shifting the burden of proof – I need not prove my claim, you must prove it is false. It is because Einstein strongly believed to the mathematics, especially his equations and he uses mathematics as proof.

Third, false equivalence – describing a situation of logical and apparent equivalence, when in fact there is none. It is found in the meaning of EFE about curvature of spacetime, that the curvature of spacetime contains density and flux of energy and momentum, or:

Curvature of spacetime = density and flux of energy and momentum.

Actually, that is highly speculative since the space and time should be in connection with astronomy, but it is shown in EFE and clear no any relationship with astronomy. And the model of spacetime really does not take into account the atmospheric medium of a massive body.(celestial bodies).

Moreover, it can't answer the question: "How is it possible for a mass to curve spacetime?"

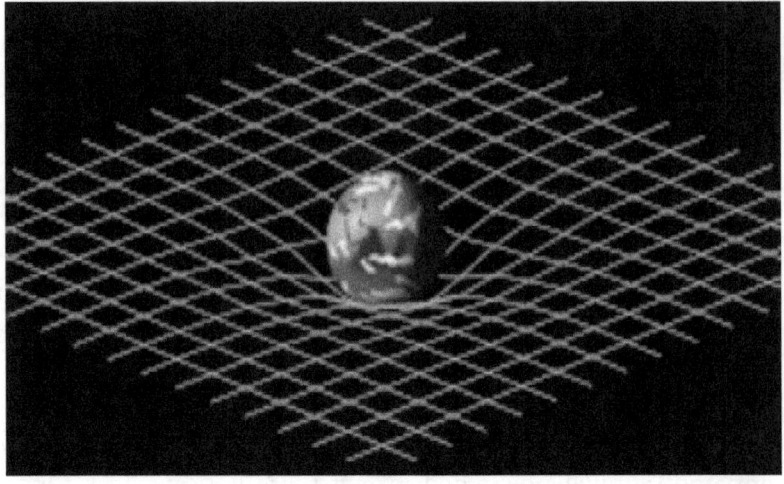

Figure 3.2: A curvature of spacetime

Fourth, general theory of relativity is about Einstein's theory of gravity, and Einstein's theory was said to "reduce" Newton's theory of gravity. To be honest, it is not to reduce Newton's theory, but refuse Newton's theory of gravity, as Einstein said gravity is not a force, but the geometry of spacetime.

In his claims, Einstein can't shows that Newton's theory of gravity is illogical or absurd. In general relativity, Einstein using mathematics as proof, but highly speculative and inaccurate.

Mathematics can be useful as a model what we see in the real world. For example, in the astronomy model of 3 D + 1 D is described in the form of celestial sphere coordinates system. This model very helpful to visualize the units and measuring in degree, minute, and arc, or in hour, minute, and seconds, and produce results that are very accurate. It's very useful.

But In general relativity, the model of 4 D is described in the form of spacetime - Einstein's field equation of gravitation. This model can not be used to visualize the units and measuring. It mean this model doesn't match up with reality.

Einstein had no idea of the units and disciplines of measurement as the goal of using mathematics in physics. So, it's useless. EFE just a formal mathematical construct with no real physical meaning. That is something logical fallacy : argumentum ad lapidem-dismissing a claim as absurd without demonstrating proof for its absurdity.

Riemann Geometry

One of the basic topics in Riemannian geometry is the study of curved surfaces in general. Riemann geometry also study higher dimensional spaces. But, there are no practical applications of Riemann geometry in astronomy. Riemann did not take an interest in the space of the astronomers. Questions about the global properties of space he cut short as "idle questions."

Projecting a sphere to a plane is very important to make a map used in astronomy. But, the astronomy doesn't using Riemann geometry. Einstein was interested to Riemann geometry, but Einstein had no idea on the basic of astronomy, therefore, he had not realized that Riemann geometry is useless in astronomy. Ironically, he declared a mathematical equations that are based on Riemann. So, Einstein's equation is also useless in astronomy, in the sense that EFE violate of astronomy. Beside that, actually, the Einstein EFE lacks a true tensor.

1.2. Testing Einstein's Hypothesis.

Our practice of observations of celestial bodies in entirely based on the fact that the sky – the Sun, Moon, Planets, and Stars – has a different appearance to observers at different points on the earth's curved surface.

In general relativity, Albert Einstein proposed test via solar eclipse. The proving method as suggested by Einstein was recorded in the book 'The Universe and Dr. Einstein', by Lincoln Barnett, published for the first time in London in June 1949. Foreword of this book was written by Albert Einstein himself.

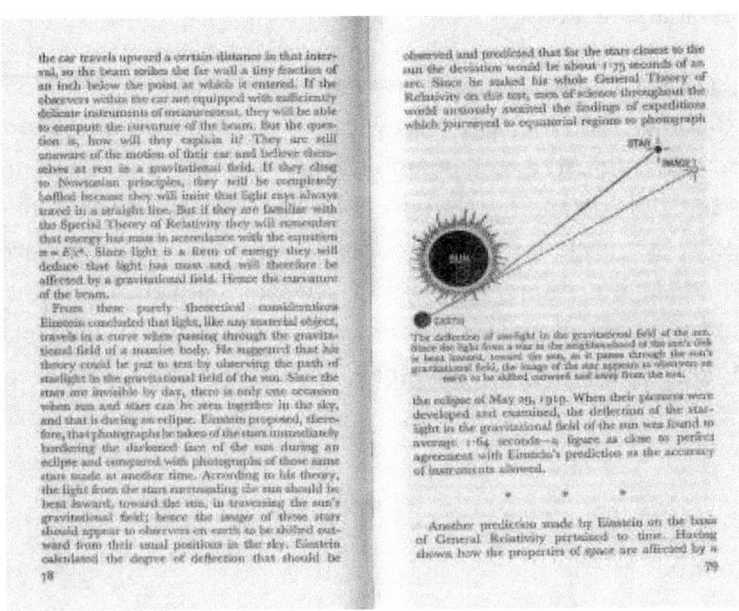

Figure 4.1: Incorrect illustration on page 79. Correct illustration if image (observed of star) looks higher than actual of star.

"From these purely theoretical considerations Einstein concluded that light, like any material object, travels in a curve when passing through the gravitational field of a massive body. He suggested that his theory could be put to test by observing the path of starlight in the gravitational field of the Sun. Since the stars are invisible by day, there is only one occasion when Sun and stars can be seen together in the sky, and that is during an eclipse.

Einstein proposed therefore, that photographs be taken of the stars immediately bordering the darkened face of the sun during an eclipse and compared with photographs of those same stars made at another time. According to his theory, the light from the stars surrounding the Sun should be bent inward, toward the Sun, in traversing the Sun's gravitational field; hence the images of these stars should appear to observer on earth to be shifted outward from their usual positions in the sky.

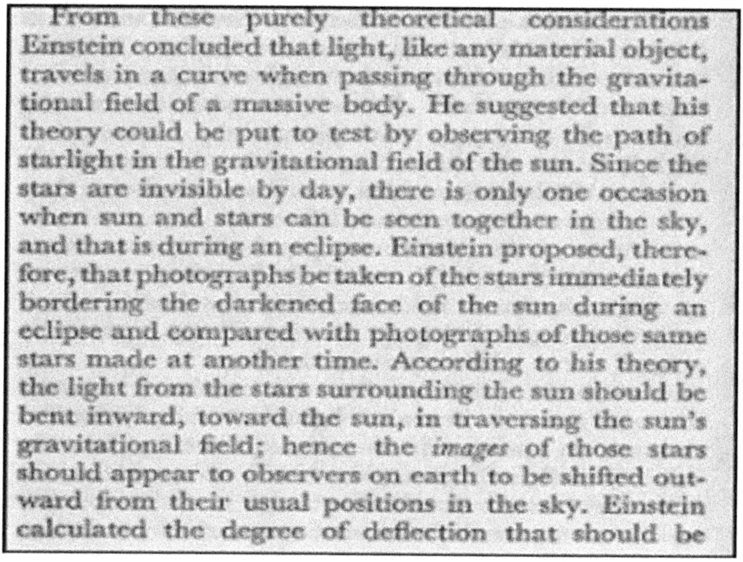

> From these purely theoretical considerations Einstein concluded that light, like any material object, travels in a curve when passing through the gravitational field of a massive body. He suggested that his theory could be put to test by observing the path of starlight in the gravitational field of the sun. Since the stars are invisible by day, there is only one occasion when sun and stars can be seen together in the sky, and that is during an eclipse. Einstein proposed, therefore, that photographs be taken of the stars immediately bordering the darkened face of the sun during an eclipse and compared with photographs of those same stars made at another time. According to his theory, the light from the stars surrounding the sun should be bent inward, toward the sun, in traversing the sun's gravitational field; hence the *images* of those stars should appear to observers on earth to be shifted outward from their usual positions in the sky. Einstein calculated the degree of deflection that should be

Figure 4.2: Einstein proposed test of general relativity.

Einstein calculated the degree of deflection that should be observed and predicted that for the stars closest to the Sun the deviation would be about 1.75 seconds of an arc.Since he staked his whole General Theory of Relativity on this test, men of science throughout the world anxiously awaited the findings of expeditions which journeyed to equatorial regions to photograph the eclipse of May 29, 1919. When their pictures were developed and examined, the deflection of the starlight in the gravitational field of the sun was found to average 1.64 seconds-a figure as close to perfect agreement with Einstein's prediction as the accuracy of instruments allowed."

Actually, the proving method of general relativity as requested by its founder, Albert Einstein, is not scientifically correct and deeply wrong. The result will always be an error:

"Einstein proposed therefore, that photographs be taken of the stars immediately bordering the darkened face of the sun during an eclipse and compared with photographs of those same stars made at another time." (Lincoln Barnett, The Universe and Dr.Einstein, London, 1949, Foreword by Albert Einstein himself, page 78).

Einstein's proving method is closely related to astronomy, especially celestial navigation.

1.Deflection of light is the difference angle between true position and apparent position of stars or the difference of altitude. In astronomy, true position and apparent position of stars are three-dimensional. But, all the photographs be taken of the stars are two-dimensional.

In this case Einstein ignored 'The Space and Time' or Celestial Sphere (Celestial Coordinate System), and ignored refraction of light as one of the fundamental concepts in astronomy.

2.All the photographs be taken of solar eclipse (the Sun and stars) are photographs of the apparent positions of the Sun and stars. The entire image as the result of refraction of light by Earth's atmosphere, and nothing to do with gravity. In this case Einstein ignored the experimental techniques

3. In astronomy, all calculations to determine the true position and the apparent position of a certain star at the sky is only applicable at a certain time and at a certain place on which such observation is performed. To compared the photographs taken during an eclipse with photographs of those same stars made at another time is not scientific.

In the view of the logical fallacy, those things are considered as the fallacy of ignoratio elenchi, or irrelevant conclusion- is indicative of misdirection in argumentation. Most of the irrelevant conclusion is caused by lack of knowledge in the field that was discussed.

Moreover, as he had written a foreword, but he had not realized there is incorrect illustration in the Lincoln Barnett's book about Albert Einstein. It's maybe a small mistake, but very annoying.

In fact, 1919 eclipse experiment-Arthur Eddington expedition-was error. Einstein's prediction the angle of bending of starlight is: 1.75 sec. arc, calculation output of Arthur Eddington as an observer from Principe Island (West of Africa) is: 1.62 sec. arc, while calculation output of Andrew Crommelin as an observer from Sobral (Brazil) is: 0.93 sec. arc).

That's why Einstein never received a Nobel prize for relativity. Einstein received a Nobel prize in 1921 for photoelectric effect, not for relativity.

"The Nobel citation reads that Einstein is honoured for "services to theoretical physics, and especially for his discovery of the law of the photoelectric effect.

At first glance, the reference to theoretical physics could have been a back door through which the committee acknowledged relativity.

However, there was a caveat stating that the award was presented "without taking into account the value that will be accorded your relativity and gravitation theories after these are confirmed in the future". [5]

Incorrect Illustration

Besides test of general relativity, we can see in the book The Universe and Dr.Einstein, page 79, something incorrect illustration. We know that Einstein had written a foreword, but he had not realized there is incorrect illustration on page 79, and it shows he does not understand about the effects of deviation of starlight in the real world, for example, when we see the stars in the sky at night.

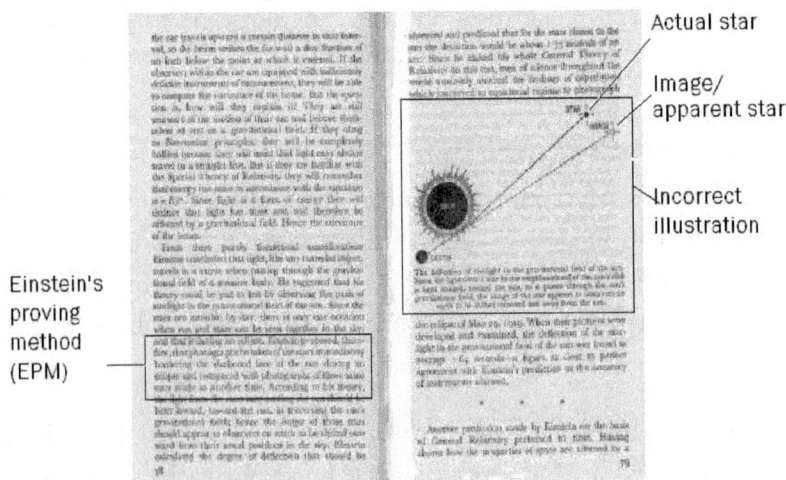

Public domain image - Archive.Org

Figure 4.3: Incorrect illustration.

Illustration in Figure 4.3 shows that image of star or apparent of star looks higher than actual of star. That is incorrect. Apparent of star always looks higher than actual of star, see in Figure 4.4.

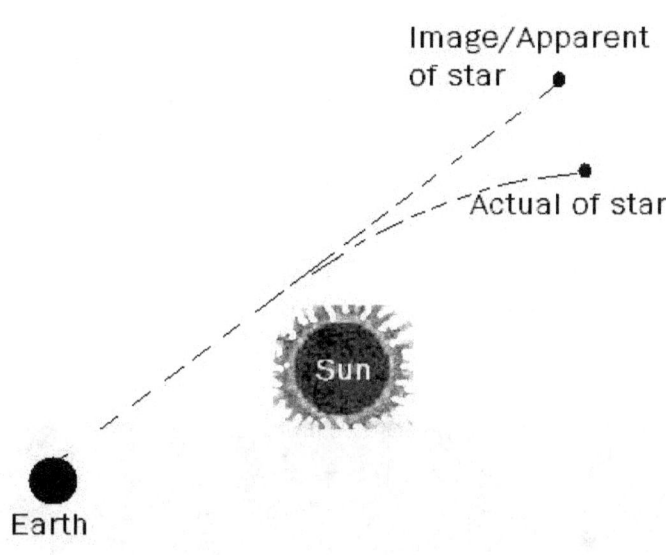

Figure 4.4.

From the two errors, the proving method and the figure /illustration, we can say that Albert Einstein had no idea on the basic fundamental of astronomy, or to be more precisely Einstein has no competence in the field of astronomy.

Actually, to be honest, Einstein general theory of relativity has been wrong since the beginning. It is not just incomplete theory, but the theory was totally wrong.

Dr.Louis Essen wrote in 1988:

"Rutherford treated it as a joke; Soddy called it a swindle ...:

....Einstein's use of a thought experiment, together with his ignorance of experimental techniques, gave a result which fooled himself and generations of scientists." (Dr.Louis Essen, inventor of atomic clock, 1988) [6]

Let us note for the general theory of relativity, if this theory was correct, then the light from stars that passed closest to the Sun would show the greatest degree of bending, and the stars whose light tracks are very far from the Sun have their lights not being bent or deflected. The stars whose lights are not deflected means that there is no difference between the image or apparent of star and the actual of star, please see in Figure 4.5.

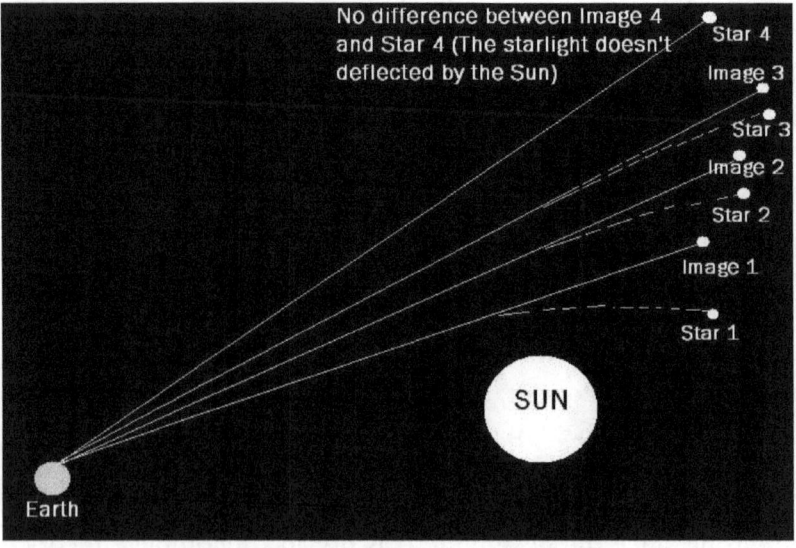

Figure 4.5

If being consistent with this theory, it means that all stars visible at night time are at the appearance of the stars on their actual position, because the said stars do not pass through the field of gravity. This is certainly incorrect if it is seen from the astronomical scientific point of view. The stars in the sky at night time and seen by the observers, all are stars on apparent position, not on their actual position.

The Reports of F.W. Dyson

Reports of F.W. Dyson was written in the document entitle "Determination of the deflection of light by the Sun's Gravitational Field, from Observations made at the Total Eclipse of May 29, 1919", published on April 27, 1920. [7]

"The results of the observations here described appear to point quite definitely to the third alternative, and confirm Einstein's generalized relativity theory. As is well- known the theory is also confirmed by the motion of the perihelion of Mercury, which exceeds the Newtonian value by 43'^ per century— an amount practically identical with that deduced from Einstein's theory.

It seems clear that the effect here found must be attributed to the sun's gravitational field and not, for example, to refraction by coronal matter. ….Clearly a density of this order is out of the question."

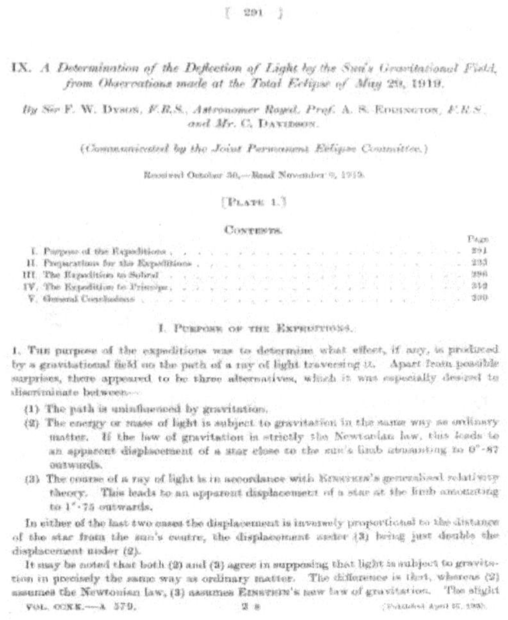

[291]

IX. *A Determination of the Deflection of Light by the Sun's Gravitational Field, from Observations made at the Total Eclipse of May 29, 1919.*

By Sir F. W. Dyson, *F.R.S., Astronomer Royal, Prof.* A. S. Eddington, *F.R.S., and Mr.* C. Davidson.

(*Communicated by the Joint Permanent Eclipse Committee.*)

Received October 30,—Read November 6, 1919.

[Plate 1.]

Contents.

I. Purpose of the Expeditions.

1. The purpose of the expeditions was to determine what effect, if any, is produced by a gravitational field on the path of a ray of light traversing it. Apart from possible surprises, there appeared to be three alternatives, which it was especially desired to discriminate between—

(1) The path is uninfluenced by gravitation.

(2) The energy or mass of light is subject to gravitation in the same way as ordinary matter. If the law of gravitation is strictly the Newtonian law, this leads to an apparent displacement of a star close to the sun's limb amounting to 0"·87 outwards.

(3) The course of a ray of light is in accordance with Einstein's generalised relativity theory. This leads to an apparent displacement of a star at the limb amounting to 1"·75 outwards.

In either of the last two cases the displacement is inversely proportional to the distance of the star from the sun's centre, the displacement under (3) being just double the displacement under (2).

It may be noted that both (2) and (3) agree in supposing that light is subject to gravitation in precisely the same way as ordinary matter. The difference is that, whereas (2) assumes the Newtonian law, (3) assumes Einstein's new law of gravitation. The slight

VOL. CCXX.—A 579. 2 S (Published April 27, 1920.)

Figure 4.6: Determination of the deflection of light by the Sun's Gravitational Field, from Observations made at the Total Eclipse of May 29, 1919 (archive.org)

According to this document, observation on 1919 eclipse take into account the effects of the refraction of light by a coronal matter - Sun's atmosphere – but in this report the effects was not found, '*Clearly a density of this order is out of the question*'

But observations in the year of 1919 eclipse without taking into account the effects of light's refraction that are caused by Earth's atmosphere, I mean the effect caused by the height of the place of observation. In the report of F.W Dyson we can't find data of the height of the place of observation, neither in Principle Island nor in Sobral. Moreover, there is no data of the correction for the altitude of the Sun and stars during eclipse. We realized, maybe at that time astronomy has not advanced as today, for example using nautical almanac if GPS fail to determine a certain position at sea.

The modern astronomy was developing after 1950. In modern astronomy, the height of observations can not be ignored, because it's in connection with height of eye. The height of eye give an effects known as terrestrial refraction.

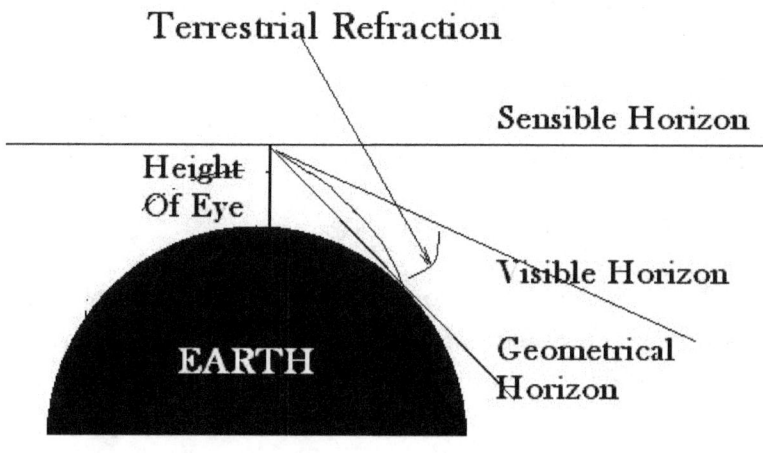

Figure 4.7: Terrestrial Refraction

This effect depend on geographyc position of observers- latitude and longitude- and not just temperature, but height of eye of an observer can't be ignored.

The most of important things is, the fundamental principle of the old astronomy is not different with modern astronomy, calculation of deviation of starlight should be made immediately, not allowed to use a comparison with the observations made at another time.

. In his report, F.W.Dyson put the figure of Einstein's precision orbit Mercury: "…is also confirmed by the motion of the perihelion of

Mercury..."

F.W. Dyson and team, including Professor Arthur Stanley Eddington seemed want to says, Einstein correct and accurate about Mercury's orbits, therefore, the 1919 eclipse experimental results is also correct.

The Perihelion of Mercury

Einstein's Mercury orbit was challenged by several scientists including Dr. Thomas Van Flandern astronomer who worked at the U.S. Naval Observatory in Washington.[8]

Excerpt of Tom Van Flandern Articles:

"Fact: The equation that accounted for Mercury's orbit had been published 17 years earlier, before relativity was invented. The author, Paul Gerber, used the assumption that gravity is not instantaneous, but propagates with the speed of light. After Einstein published his general-relativity derivation, arriving at the same equation, Gerber's article was reprinted in *Annalen der Physik* (the journal that had published Einstein's relativity papers). The editors felt that Einstein should have acknowledged Gerber's priority. Although Einstein said he had been in the dark, it was pointed out that Gerber's formula had been published in Mach's Science of Mechanics, a book that Einstein was known to have studied. So how did they both arrive at the same formula?

Tom Van Flandern was convinced that Gerber's assumption (gravity propagates with the speed of light) was wrong. So he studied the question. He points out that the formula in question is well known in celestial mechanics. Consequently, it could be used as a "target" for calculations that were intended to arrive at it. He saw that Gerber's method "made no sense, in terms of the principles of celestial mechanics." Einstein had also said (in a 1920 newspaper article) that

Gerber's derivation was "wrong through and through."

So how did Einstein get the same formula? Van Flandern went through his calculations, and found to his amazement that they had "three separate contributions to the perihelion; two of which add, and one of which cancels part of the other two; and you wind up with just the right multiplier." So he asked a colleague at the University of Maryland, who as a young man had overlapped with Einstein at Princeton's Institute for Advanced Study, how in his opinion Einstein had arrived at the correct multiplier. This man said it was his impression that, "knowing the answer," Einstein had *"jiggered the arguments until they came out with the right value."*

In the article entitle The Perihelion Precession of Mercury, Miles Mathis wrote:

"Einstein assigns his famous number .45 to precession per year while having no mathematical or theoretical reasons for that time assignment. By checking all his famous papers on GR, we find that he certainly found the number .45, but we find nothing in his equations that makes that per year. He simply assumed the period of precession, since his number matched historical equations." [9]

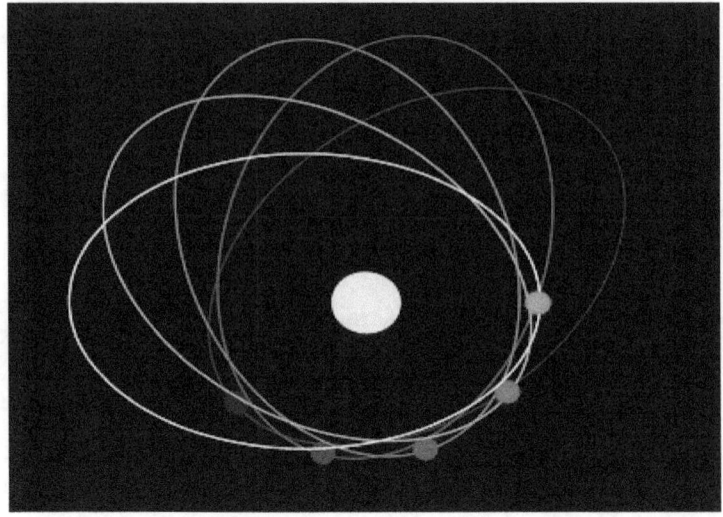

Figure 4.8: The perihelion precession of Mercury (Wikipedia)

Last and not least, the report of F.W Dyson give a confirmation: "The course of a ray of light is in accordance with Einstein's generalized relativity theory. This leads to an apparent displacement of a star at the limb amounting to 1.75 sec.arc outwards."

Deflection of light by the Sun

Wikipedia informs us about Tests of General Relativity, actually Albert Einstein proposed three tests of general relativity, subsequently called the classical tests of general relativity, in 1916:

1. the perihelion precession of Mercury's orbit

2. the deflection of light by the Sun

3. the gravitational redshift of light

In the letter to the London Times on November 28, 1919, he described the theory of relativity and thanked his English colleagues for their understanding and testing of his work. He also mentioned three classical tests with comments:

"The chief attraction of the theory lies in its logical completeness. If a single one of the conclusions drawn from it proves wrong, it must be given up; to modify it without destroying the whole structure seems to be impossible."

About the second test, the deflection of light by the Sun, there is the test has two meanings. First, deflection of light by the field of gravity around the Sun without effects of Earth's atmosphere. In this meaning, the test must be carry out from spacecraft. Second, deflection of light by the Sun as seen from Earth. See in Figure 4.9.

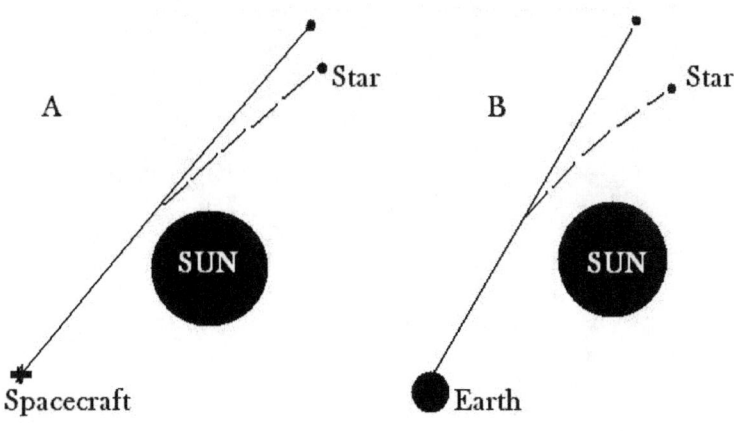

Figure 4.9: Two meanings deflection of light by the Sun

As we know, Einstein proposed test of general relativity via eclipse, by using photograph, and in this test Einstein had ignored refraction of light, it's mean the test is the same with A on Figure 4.9. That, in accordance with F.W.Dyson's document that says:" It seems clear that the effect here found must be attributed to the sun's gravitational field and not, for example, to refraction by coronal matter."

But, if we read how Eddington demonstrated that Einstein was right, illustrated on London News, 22 November 1919,the illustration of deflection of light by the Sun is the same with the meaning of B on Figure 4.9, the deflection of light as seen from the Earth.

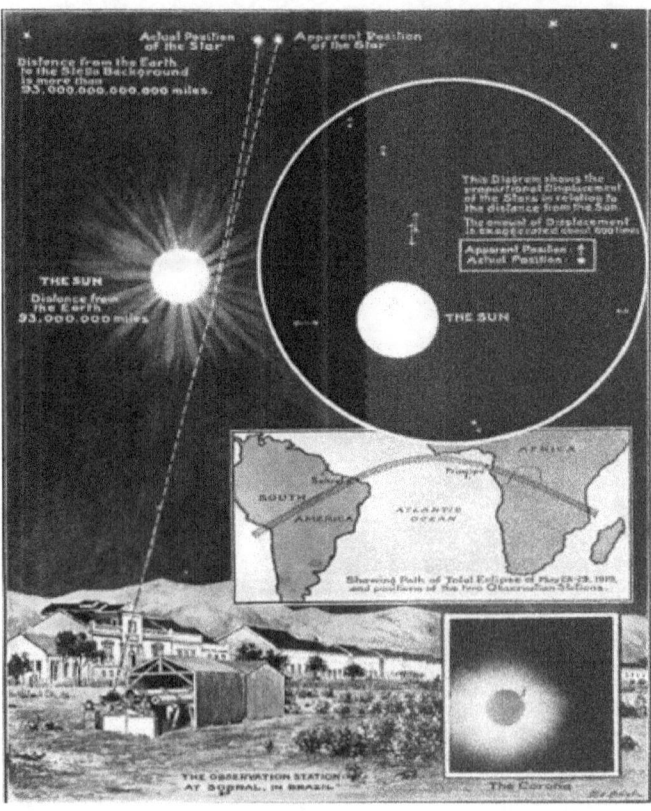

Figure 4.10: Deflection of light by the Sun-Images credit: New York Times, 10 November 1919 (L); Illustrated London News, 22 November 1919 (R).

From the above discussion we can drew a conclude, there are something logical fallacies - the fallacy of ignoratio elenchi, or irrelevant conclusion. They want to measure bending of light by the Sun, but they measured bending of light by Earth's atmosphere using Einstein proving method via eclipse. Moreover, this method is not scientifically correct and deeply wrong. Note that "the deflection of light by the Sun ", labeled A in Figure 4.9, as seen from Earth is not even measurable.

But, it doesn't matter what the meanings of deflection of light by the Sun, A or B on Figure 4.9. As we know the Nobel Committee in 1921 didn't agree and make a note for Einstein: "without taking into account the value that will be accorded your relativity and gravitation theories after these are confirmed in the future".

Invalidity of Einstein's Hypothesis

Einstein's Special and General theory of relativity is widely known as theories in modern physics, as they said, that are pass of every tests for more than 100 years. Stephen Hawking said: "General Relativity was a major intellectual revolution that has transformed the way we think about the universe. It is a theory not only of curved space, but of curved or warped time as well." This is in fact what happens repeatedly any time.

They had not realized, or maybe they know but they know not if they know, or they know not if they know not, that general relativity was totally wrong.

As we know, Albert Einstein proposed test of general relativity only three classical tests, as it has been discussed, and he states:

"The chief attraction of the theory lies in its logical completeness. If a single one of the conclusions drawn from it proves wrong, it must be given up; to modify it without destroying the whole structure seems to be impossible."

He stressed that his theory lies in its logical completeness. Facts, there are many of logical fallacies of Einstein. The first experiment in 1919 is very clear the result was error, Einstein's proving method is not scientifically correct, but they don't admit that they are wrong, The eclipse experiment was repeated in the year of 1922, and then in the year of 1953, and again in the year of 1973. All the eclipse experiments with results Einstein's general relativity was right.

Here a big question: are they had no idea on the basic concepts of astronomy?

Facts, through the search via internet we can't find any documents of eclipse experiment in the year 1922, 1953, and 1973. So, is it a swindle as Sodhi said?

But the experiments never ended, although Einstein had said: "If a single one of the conclusions drawn from it proves wrong, it must be given up; to modify it without destroying the whole structure seems to be impossible."

We should be back to the basic concepts of the scientific method.

The Scientific Method.

"The scientific method is a process for creating models of the natural world that can be verified experimentally. The scientific method requires making observations, recording data, and analyzing data in a form that can be duplicated by other scientists. In addition, the scientific method uses inductive reasoning and deductive reasoning to try to produce useful and reliable models of nature and natural phenomena.

Inductive reasoning is the examination of specific instances to develop a general hypothesis or theory, whereas deductive reasoning is the use of a theory to explain specific results. In 1637 René Descartes published his Discours de la Méthode in which he described systematic rules for determining what is true, thereby establishing the principles of the scientific method." [10]

The scientific method has four steps:

1. Observation and description of a phenomenon (a concept),

2. Formulation of a hypothesis to explain the phenomenon,

3. Test the hypothesis. If experiments do not confirm the hypothesis, the hypothesis must be rejected or modified (Go back to Step 2),

4. Establish a theory based on repeated verification of the results.

See in Figure 5.1

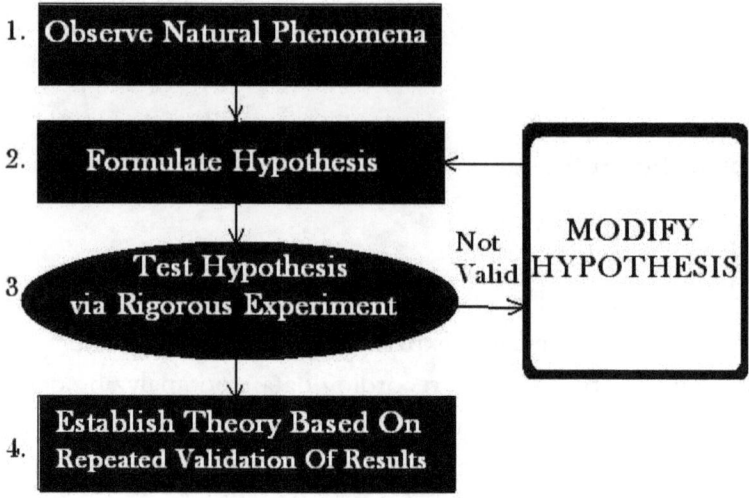

Figure 5.1: Four steps of scientific method.

In this case of Einstein's hypothesis, step 1 observe natural phenomena of general relativity is constructed by thought experiments and his mathematics. Step 2 formulate hypothesis, Einstein concluded:

"From these purely theoretical considerations Einstein concluded that light, like any material object, travels in a curve when passing through the gravitational field of a massive body. Einstein calculated the degree of deflection that should be observed and predicted that for the stars closest to the Sun the deviation would be about 1.75 seconds of an arc"

.Einstein's prediction of 1.75 sec.arc without taking into account the altitude of the star/Sun as the object of observation. This is a fatal mistake because this prediction has no scientific meaning in astronomy. The value bending of starlight isn't the same at the minimum solar eclipse and at the maximum solar eclipse. The important thing to be note, in astronomy, deviation or bending of starlight will always vary depending on the altitude of the object of observation. In this case, hypothesis Einstein is not valid. Einstein hypothesis doesn't meet requirements of scientific method.

The above hypothesis raise a question, where does the observation is made, from the space by a spacecraft or from the Earth? Einstein explains that observation is made from Earth:

"Einstein proposed therefore, that photographs be taken of the stars immediately bordering the darkened face of the sun during an eclipse and compared with photographs of those same stars made at another time. According to his theory, the light from the stars surrounding the Sun should be bent inward, toward the Sun, in traversing the Sun's gravitational field; hence the images of these stars should appear to observer on earth to be shifted outward from their usual positions in the sky"

From the above statement we drew a conclude there are two points:

1.Observation is made from the Earth, and not from space

2.Test method during a solar eclipse.

If observation is made from the Earth, the results is not deflection of light by the Sun (a curve space around the Sun), but deflection of light by the Earth's atmosphere. See in Figure 5.2.

According to Einstein, light travels in a curve when passing through the gravitational field of a massive body. In the Figure 5.2, starlight travels in a curve when passing through the gravitational field of the Sun, it labeled as x: deflection of light by the Sun. And starlight travels in a curve when passing through the gravitational field of the Earth, it labeled as y:deflection of light by the Earth's atmosphere. From the Figure 5.2 it's clear that Einstein's hypothesis is not valid.

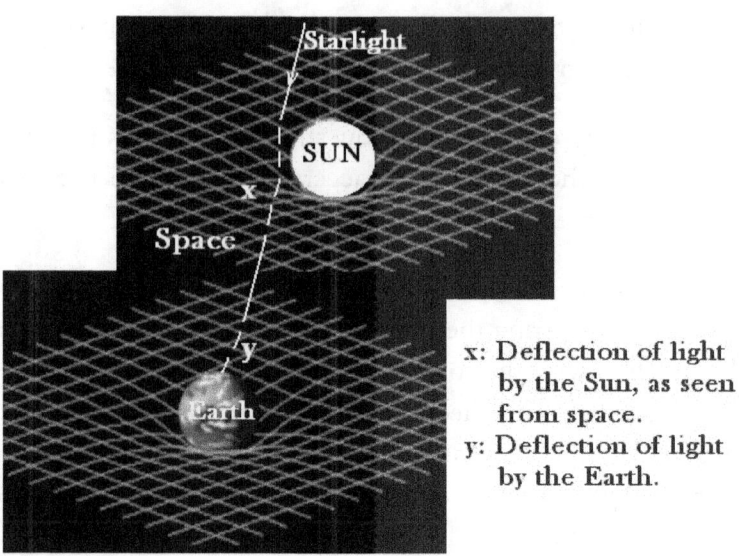

Figure 5.2: Invalidity Einstein's hypothesis of general relativity.

And if consistent with the principles of the scientific method, the hypothesis could not proceed to the step 3.test hypothesis via rigorous experiment. The hypothesis is invalid and must be rejected or modified. Moreover, it is known that the test method proposed by Einstein is not scientifically correct.

Einstein said: "If a single one of the conclusions drawn from it proves wrong, it must be given up; to modify it without destroying the whole structure seems to be impossible."

Thus, it is clear that general relativity was totally wrong since the beginning.

Have you ever heard or reading, they says that general relativity pass of every tests?

It doesn't matter what they says, we have a principles of the scientific method and we have a rule, something like this: " If a person as theory founder conveys a theory and at the same time shows its proving method, however after being tested by expert it is found out that his proposed proving method is proven to be unable to be performed due to not being scientific, then automatically such proposed theory prematurely falls by itself. The proving cannot be carried out by other methods not as requested by the theory founder, since it is reasonably assumed that such proving is made based on belief."

In his book A Brief History Of Time, Stephen Hawking wrote:" What should you do when you find you have made a mistake like that? Some people never admit that they are wrong and continue to find new, and often mutually inconsistent, arguments to support their case…"

This is in fact happens repeatedly for more than 100 years, moreover, after Einstein passed away in 1955.

Shapiro time delay effect

According to Wikipedia, the Shapiro time delay effect, or gravitational time delay effect, is one of the four classic solar system tests of general relativity. Radar signals passing near a massive object take slightly longer to travel to a target and longer to return than they would if the mass of the object were not present. The time delay is caused by the slowing passage of light as it moves over a finite distance through a change in gravitational potential.

The first tests, performed in 1966 and 1967 using the MIT

Haystack radar antenna, were successful, matching the predicted amount of time delay. The experiments have been repeated many times since then, with increasing accuracy.

Here are the things I want to know more. Was the Shapiro experiment has taken into account the effects of the refraction of light? Because, actually, Shapiro time delay effect can be explained without general relativity assumption of the geometry change in gravitational fields. Time delay of light can be explained by understanding the effects of refraction. See in the previous discussion in the subtitle 'A revelation of Universal law'.

Hafele-Keating experiment

Agaiin, Wikipedia makes familiar to us, the Hafele–Keating experiment was a test of the theory of relativity. In October 1971, Joseph C. Hafele, a physicist, and Richard E. Keating, an astronomer, took four cesium-beam atomic clocks aboard commercial airliners. They flew twice around the world, first eastward, then westward, and compared the clocks against others that remained at the United States Naval Observatory. When reunited, the three sets of clocks were found to disagree with one another, and their differences were consistent with the predictions of special and general relativity.

General relativity predicts an additional effect, in which an increase in gravitational potential due to altitude speeds the clocks up. That is, clocks at higher altitude tick faster than clocks on Earth's surface. This effect has been confirmed in many tests of general relativity, such as the Pound–Rebka experiment and Gravity Probe A. In the Hafele–Keating experiment, there was a slight increase in gravitational potential due to altitude that tended to speed the clocks back up. Since the aircraft flew at roughly the same altitude in both directions, this effect was approximately the same for the two planes, but nevertheless it caused a difference in comparison to the clocks on the ground.

This is the same as Shapiro experiments, the experiments can be explained without assumption of the time dilation of Einstein's relativity. Facts, density of Earth's atmosphere, temperature, and pressure at height of the plane give an effect to clocks. Clocks at higher altitude tick faster than clocks on Earth's surface. It is not caused by gravity, but by air density of atmosphere. Closer to the earth surface, the air is denser compared to the density of the air layer above it. The density is getting looser or weaker when it is getting higher.

It is has been known in traveling on an airplane. At higher altitude the density of amosphere is getting looser or weaker, and less of friction on an airplane. Traveling in weaker density of atmosphere an airplane can move faster than in denser atmosphere.

Gravitational Redshift, Gravitational Lensing, Gravitational wave, Speed of gravity, Gravity Probe B and GPS.

As we know, there are many of hypotheses or theories in modern physics that are associated to general relativity: gravitational redshifts, gravitational lensing, gravitational wave, speed of gravity, and about GPS. If general relativity was correct, those theories can be true or untrue. But if general relativity was wrong, all of the hypotheses or theories can not be true.

For example in the case of gravitational redshift and gravitational lensing, these are not supported any evidence, just a possibility and another possibility. Gavitational redshift and gravitational lensing can be explained without Einstein's curvature of space.

"Many of the general relativity tests such as bending of light near a star and gravitational red/blue shift are explained without general relativity & without Newtonian approach. The author first casts doubts on both, the Newtonian and the relativistic approach; and proposes a novel alternative explanation. The new alternative explanation is based on refraction phenomenon of optics. It predicts that as the ray passes through/near the star's atmospheric medium, it bends due to refraction phenomenon towards star core, like a ray bends while passing through a prism or water drop. A semi empirical estimation of the atmospheric height h and its refractive index µ are made to find the refraction results. The refraction based theory also suggests new explanation for gravitational red/blue shift; it tells that frequency ? remains constant (as it is so in refraction phenomenon) and the red/blue shift is due to change in wavelength ? due to change in velocity of light c in the medium.

Estimated results for bending of light and the red/blue shift etc. with the new approach though agree well with known values, but important thing is that the physics is quite different. Also discussed are black hole, gravitational lensing and space time in the new perspective of refraction. The proposed refraction based theory proposes a new look on black hole, suggesting that black hole formation is critically due to total internal reflection within atmosphere and subsequent absorption into the star core. Gravitational lensing is explained as real refraction lensing with possibility of chromatic aberration.

The new refraction based theory also makes a few new predictions. The present paper also suggests a possible alternative to the Einstein's curved geometry of spacetime, and indicates that the fabric of spacetime which wraps(curves) around the mass is not the empty vacuum but the atmospheric medium. The new refraction based approach providing alternative to general relativity, could have important bearing on understanding of spacetime, gravity and cosmology !" (Professor R. C. Gupta, India, on his paper 'Bending of Light Near a Star and Gravitational Red/Blue Shift: Alternative Explanation Based on Refraction of Light') [11]

And about gravity probe B, it is something mission impossible. They take measurements in the Earth's atmosphere, thermosphere, not in space, of course, the result is very doubtful. Obviously, it was not possible.

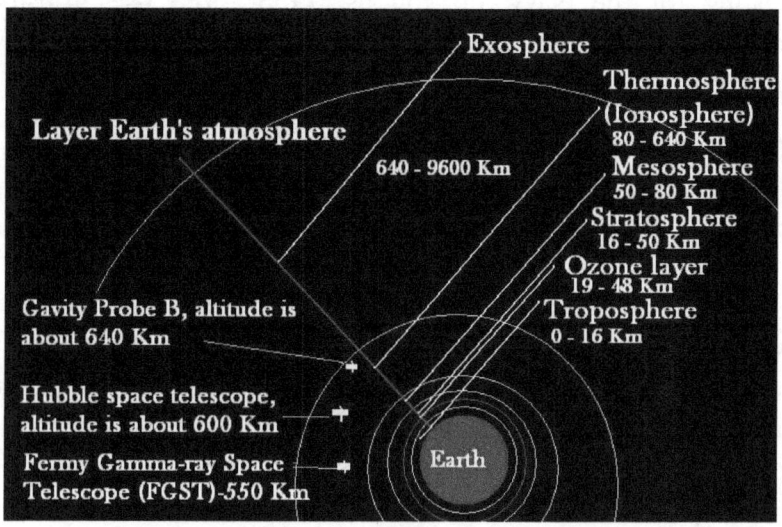

Figure 5.3: The Gravity Probe B (GP-B) based on observations of spinning gyroscopes onboard a satellite in a near-polar, near-circular orbit at an altitude of about 640 km around the Earth (ScienceDirect).

Mile Mathis wrote in his website:

"Gravity Probe B project that was given a grade of F in a NASA review in 2008 by a group of senior advisors, and denied any further funding since, "the reduction in noise needed to test rigorously for a deviation from general relativity is so large that any effort ultimately detected by this experiment will have to overcome considerable (and in our opinion, well justified) scepticism in the scientific community". They continue:

The noisy data meant that GPB could not measure the effects as precisely as astronomers had by firing laser beams at mirrors left of the Moon by the Apollo astronauts."[12]

Someone ask to me,: "With all due respect, when you are navigating for the Navy, do you use GPS? GPS uses Einstein's equations of General Relativity to get the precision required:). And would be useless without them .."

Actually, GPS doen't use and doesn't need Einstein's relativity. This is confirmed by Professor Tom Van Flandern in article 'The Speed of Gravity: Why Einstein Was Wrong and Newton Was Right':

"Van Flandern goes on to discuss GPS clocks, which are often cited as being proof positive of Einstein's relativity. It may surprise you, but the GPS system doesn't actually use Einstein's field equations. In fact, this paper by the U.S. Naval Observatory tells us that, while incorporating Einstein's equations into the system may slightly improve accuracy, the system itself doesn't rely on them at all. To quote the opening line of the paper, "The Operational Control System (OCS) of the Global Positioning System (GPS) does not include the rigorous transformations between coordinate systems that Einstein's general theory of relativity would seem to require."

Van Flandern explains why this is so:

"Finally, the Global Positioning System (GPS) showed the remarkable fact that all atomic clocks on board orbiting satellites moving at high speeds in different directions could be simultaneously and continuously synchronized with each other and with all ground clocks. No "relativity of simultaneity" corrections, as required by SR, were needed. This too seemed initially to falsify SR. But on further inspection, continually changing synchronization corrections for each clock exist such that the predictions of SR are fulfilled for any local co-moving frame. To avoid the embarrassment of that complexity, GPS analysis is now done exclusively in the Earth-centered inertial frame (the local gravity field). And the pre-launch adjustment of clock rates to compensate for relativistic effects then hides the fact that all orbiting satellite clocks would be seen to tick slower than ground clocks if not rate-compensated for their orbital motion, and that no reciprocity would exist when satellites view ground clocks."

Einstein general theory of relativity look like the best actor in modern physics for more than 100 years, but in facts, many of the general relativity tests can be explained without Einstein geometry of spacetime.

1.3. Astronomical Test of General Relativity

In the previous discussion it has been explained that the test of general relativity proposed by Einstein is not scientifically correct, and the result will always be an error.

Dr. Louis Essen, a physicist and well-known as the inventor of the atomic clock, has written several articles express criticism to the relativity. Louis Essen said that Einstein use of a thought experiment, together with his ignorance of experimental techniques. Therefore, Einstein made mistake in his test and has been said as the fallacy of ignoratio elenchi or irrelevant conclusion. It is because Einstein had no idea in the field of astronomy.

Now, we need to know how test should be done in the scientific views of astronomy or the astronomical method to test general relativity. Astronomical tests of general relativity through the event of a total solar eclipse, really easy to do and not costly.

Keep in mind that the test to be performed related to the calculation of the deviation of starlight, and at the same time we want to know the altitude of the Sun/Moon during the eclipse. Therefore, it is necessary to use a sextant and nautical almanac, and not using the camera. This test is direct observation during a solar eclipse, nothing the observation 'is made at another time'.

The astronomical method has three steps:

1. Determine the place of observation to make sure the height of eye of an observer from the sea level.. If observations were made from the beach, height of eye is about 2 - 2.5 meters. If the observations were made from the top of the of a building at the beach or from the ships, high of eye can be more than 10 meters

2. Measuring the apparent altitude of the celestial bodies (stars, Sun, Moon, or planets) using the sextant, and noted the time. Once you

have taken a sight and noted the time in local time, you'll need to know the deviation of the celestial bodies.

3.Using the Nautical Almanac, find the deviation of stars, Sun, Moon, or planets. With data of the apparent altitude of star and height of eye of an observer, deviation of starlight can be calculated.

For example, if the test was made during a solar eclipse in 1919, then using a nautical almanac of 1919. If the test is.made during a solar eclipse in 2017, then using a nautical almacac of 2017.

The Sextant

"There's nothing mystical or complicated about a sextant. All it is a device that measures the angle between two objects." (John P.Budlong)

Wikipedia informs us about the sextant: " A sextant is a doubly reflecting navigation instrument that measures the angle between two visible objects. The primary use of a sextant is to measure the angle between an astronomical object and the horizon for the purposes of celestial navigation. The estimation of this angle, the altitude, is known as sighting or shooting the object, or taking a sight. The angle, and the time when it was measured, can be used to calculate a position line on a nautical or aeronautical chart -- for example, sighting the Sun at noon or Polaris at night (in the Northern Hemisphere) to estimate latitude.

Sighting the height of a landmark can give a measure of distance off and, held horizontally, a sextant can measure angles between objects for a position on a chart. A sextant can also be used to measure the lunar distance between the moon and another celestial object (such as a star or planet) in order to determine Greenwich Mean Time and hence longitude. The principle of the instrument was first implemented around 1730 by John Hadley (1682–1744) and Thomas

Godfrey (1704–1749), but it was also found later in the unpublished writings of Isaac Newton (1643–1727).

In the first time you see a sextant, you might think that a sextant looks pretty complicated, but it really isn't. There are only three basic parts, as shown in Figure 5.1.

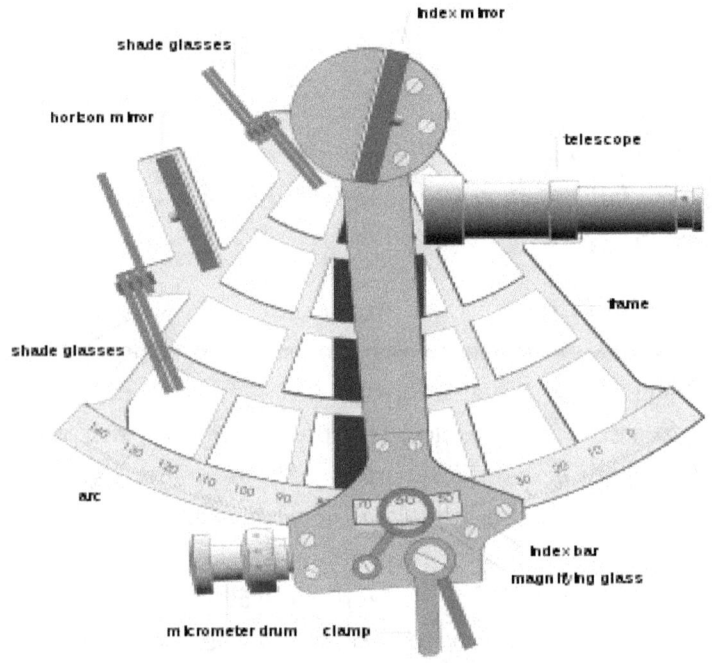

Figure 5.1: Sextant (Wikipedia)

A fixed horizon glass and an adjustable index mirror, enabling you to see the Sun or star and the horizon at the same time, and scale (arc) to read the star altitude. Some refinement are usually added, such as a sighting tube or telescope, and shades to tame the Sun's brightness and to cut glare from the horizon.

Figure 5.2: Using a Sextant (Wikipedia)

Taking a sight

The basic concept is to measure the angle from the Horizon to the Sun, or Star, Moon, and planets.

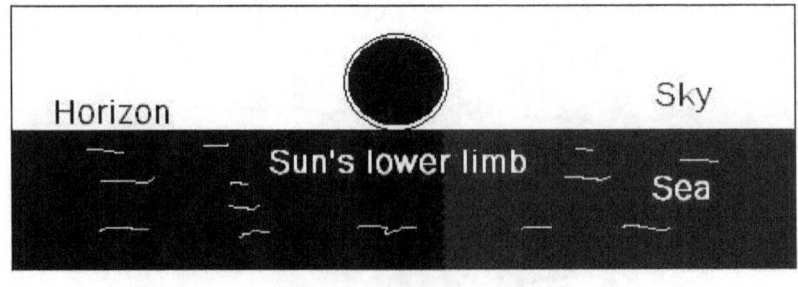

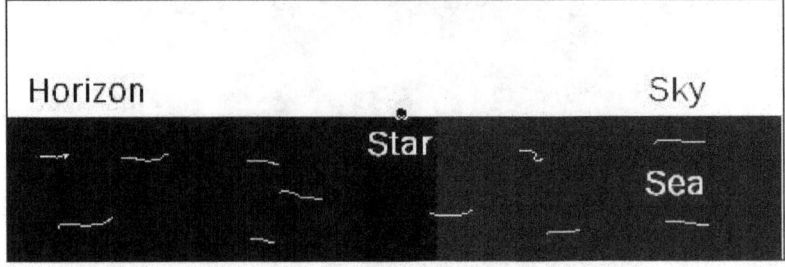

Figure 5.3: The Sun and Star image seen through a sextant

Reading The Scale

The main scale of sextant is the arc. It is to read degrees directly. For deviding the degree into minutes of angle, a vernier is provided. In the Figure 5.4, measuring the altitude of a star and the angle is 61 degrees and 30 minutes.

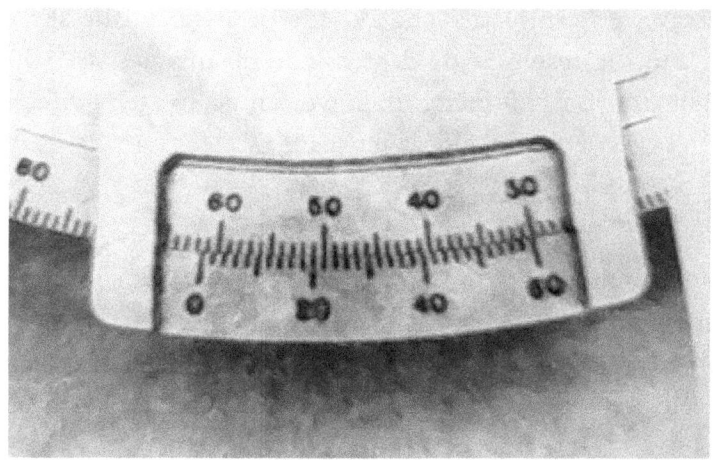

Figure 5.4: A tangent vernier

There is another type of a tangent vernier in the Figure 5.5, measuring the altitude of a star and the angle is 40 degrees and 51 minutes.

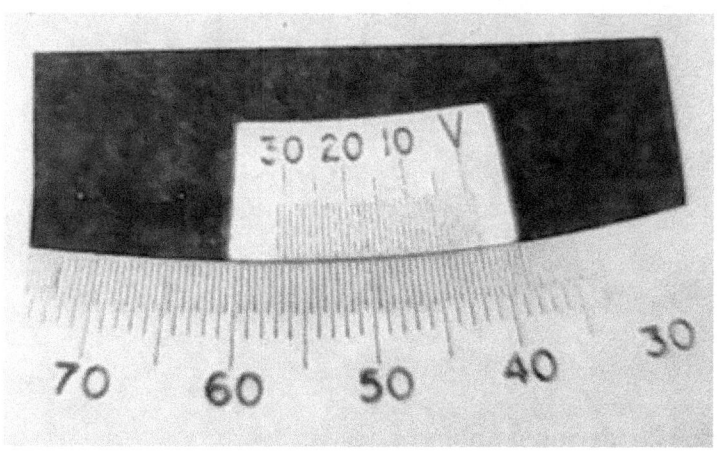

Figure 5.5

Let's say we've measured the altitude of a star in the sky using a sextant, and the results is 61 degrees and 30 minutes (61^0 30'), and the height of eye is 2,5 meter, then we can go to steps three to find deviation of starlight in nautical almanac.

Nautical Almanac

A nautical almanac is a publication describing the positions of a selection of celestial bodies. The Almanac specifies for each whole hour of the year the position on the Earth's surface (in declination and Greenwich hour angle) at which the sun, moon, planets and first point of Aries is directly overhead. (Wikipedia).

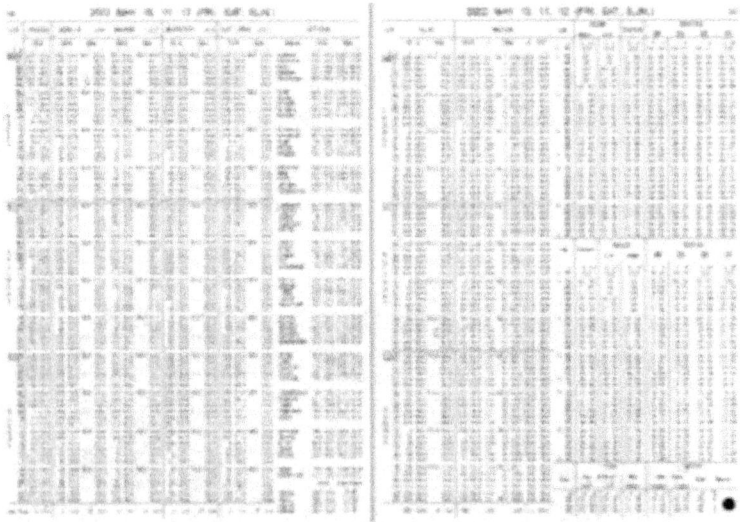

Figure 5.6: Two sample pages of nautical almanac.

Altitude Correction Table

A table of Apparent Correction and DIP is given on the inside front cover of the nautical almanac, under the heading 'Stars and Planets' and DIP. DIP is correction for the height of eye of an observer from the sea level or Height of Eye (Ht).

Actually, App.Corr and DIP are correction of refraction: astronomical refraction and terrestrial refraction. Correction in minute and tenth, and always subtracted. See Figure 5.7.

ALTITUDE CORRECTION TABLE

Stars and Planets		DIP			
App.Corr.	App.Corr.	Ht	Corr	Ht	Corr.
9.55 -5.3 10.7		2.4 -2.8 2.6		Venus 0 +0.1 60	
60.26 -0.5 65.06		18.4 -7.6 18.8		Mars 0 +0.1 60	

Figure 5.7: Part of App.Corr and DIP of Nautical Almanac of the year 2017.

Apparent Altitude of star: $61°30'$

App.Corr   :   - 0.5 minutes.

DIP Corr   :   - 2.8 minutes

Deviation of starlight = App.Corr + DIP Corr.

$$= - 0.5 - 2.8$$

$$= - 3.3 \text{ minutes.}$$

Actual altitude of star : $61°30' - 3.3' = 61°26.7'$

From the previous discussion about spacetime, Einstein's mathematics, and Einstein's proving method we know that Einstein had no idea of the basic of astronomy as Dr.Louis Essen said " ignorance of experimental techniques'. Einstein made assumption of spacetime in his mathematic not because he could prove it, but because no one disprove it

Now we know how test of general relativity should be made . And we can apply astronomical method at the 2017 solar eclipse, Monday, August 21, in USA.

Test General Relativity on the 2017 Solar Eclipse.

"…during totality on August 21, 2017—although it will be close to midday—you'll easily be able to see 4 planets with the unaided eye near the eclipsed sun!

In order of brightness, these planets will be Venus, Jupiter, Mars and Mercury. Mars is slightly brighter than Mercury, but so nearly the same in brightness that you probably won't notice a difference."(Eddie Irizarry, See 4 planets during 2017 solar eclipse, earthsky.org).

On the 2017 solar eclipse, Monday, August 21, in USA, the bright star Regulus make a thrilling sight shinning near the Sun's corona, this event is the best chance to test or re-examine Einstein's general theory of relativity.

As we know, at his theory of general relativity, Einstein declared a new law on gravity, stating that gravity was not a force as commonly known at the Newton's gravity theory, but a part of inertia. His gravity law illustrated the object behaviour at the gravity field, for instance the planets, not in the sense of 'the attracting forcer' but only in the sense of the crossing track being taken.

For Einstein, gravity is a part of inertia. The movement of stars and planets originates from their inertia derivation, and the crossing track taken is determined by the space metrical nature, or more precisely the continuous space-time.

Einstein concluded that the light just as other material objects, moved in curve if gravity field of an object was massive. Einstein suggested that his hypothesis could be tested to observe the crossing track of the star light at gravity field of the sun. Due to the fact that the stars are not visible at day time, there is only one chance when the sun and the stars can be seen together at the sky, and that is the time when there is a solar eclipse.

Special phenomena of 2017 solar eclipse, that we will easily be able to see 4 planets, and the bright star Regulus make a thrilling sight shinning near the Sun's corona.

Apparent position of the star Regulus during total solar eclipse meet the conditions in accordance with Einstein hypothesis

According to Einstein's hypothesis, the star light visible around the sun would be bent inwards, toward the sun at the time when passing through the gravity field of the sun. Einstein calculated the level of their deviation and predicted that for the stars observed being the closest to the Sun, their deviation was about 1.75 seconds of an arc.

Test of general relativity

In this test, we use terminology "deflection of light by the Sun" as Einstein's idea on general relativity, and Einstein proposed test via eclipse. In this book, "tests" is made by using astronomical data that has been predicted at the 2017 total solar eclipse, in USA, August 21.

As explained in previous discussion, there are three steps to test general relativity using astronomical method: determine the place of observation, measuring the apparent altitude of star using the sextant, and apply data to find the deviation of starlight in the Nautical Almanac.of 2017.

To get the most accurate results, the test should be carried out at the locations of a total eclipse, in this case in the USA, on August 21,

2017.

According to Earthsky Website, any location along the path of totality from Oregon to South Carolina can enjoy good weather on eclipse day, but the western half of the United States, especially from the Willamette Valley of Oregon to the Nebraska Sandhills, will enjoy the very best weather odds. Therefore, the best place of direct observation to test general relativity is in Oregon, sea waters near the Oregon coast or on the beach of Madras.

In this case, the measurement using a sextant. The sextant allows us to very accurately measure the apparent altitude of stars, it's because there are visible horizon from the sea or beach. According to Earthsky Website, totality begins at 10:19 a.m, and duration of totality is 2 minutes, 4 seconds.

Prediction at 2017 total solar eclipse.

In this book, prediction is made based on astronomical data from NASA Eclipse Website, the Astroadventure Website, and Earthsky Website.

Event	Date	Time (UT)	Alt	Azi
Start of partial eclipse (C1) :	2017/08/21	16:05:33.3	28.1°	101.6°
Start of total eclipse (C2) :	2017/08/21	17:17:42.7	40.2°	117.2°
Maximum eclipse :	2017/08/21	17:18:42.1	40.4°	117.5°
End of total eclipse (C3) :	2017/08/21	17:19:41.7	40.5°	117.7°
End of partial eclipse (C4) :	2017/08/21	18:38:33.0	51.4°	140.9°

Figure 7.1: 2017 eclipse over Oregon (NASA Eclipse Website)

Apparent altitude of 2017 eclipse over Madras, Oregon

Sun/Moon : 41.6 ^{0}
Mercury : 31.2^0
Regulus : 40.8^0
Mars : 49.1^0
Venus : 65.3^0

Table 1. Planets & Stars in the Eclipse Sky at Time of Total Eclipse Over Madras

Object	Magnitude	Altitude	Azimuth
Sun/Moon	-	41.6°	119°
Mercury	+3.3	31.2°	118°
Venus	-4.0	65.3°	163°
Mars	+1.8	49.1°	124°
Jupiter	-1.8	-3.3°	95°
Saturn	+0.4	-57°	55°
Regulus	+1.4	40.8°	118°

Figure 7.2: 2017 eclipse ove Madras (Earthsky Website)

In Figure 7.2, we can see the altitude of Regulus is 40,8 deggrees or 40 degrees and 48 minutes (40^{0}48'). It is apparent altitude of

Regulus.

Imagine you measuring the altitude of Regulus star in the sky at night using sextant, from the beach and height of eye is about 2,5 meters, then reading the scale and you got 40,8 degrees. Taking sight of one star in the sky normally take the time no more than 1 - 2 minutes.

Then, to find deviation of Regulus-or bending of light-we can see on Table 1 Nautical Almanac of 2017, Altitude Correction and DIP for Star and Planet. Remember, you've data of height of eye is 2,5 meters, and apparent altitude of Regulus is 40,8 degrees. See Figure 7.2.

ALTITUDE CORRECTION TABLE

Stars and Planets		DIP			
App.Corr.	App.Corr.	Ht	Corr	Ht	Corr.
40.08 42.44 (-1,1)		2.4 2.6	(-2.8)	Venus 0 +0.1 60	
60.26 -0.5 65.06		18.4 18.8	-7.6	Mars 0 +0.1 60	

Figure 7.3

You can see in Figure 7.3, correction of height of eye for 2,5 meters is about -2,8 minutes, and correction of apparent altitude of Regulus for 40,8 degrees is about – 1,1 minutes. Then, we find deviation of Regulus is – 2,8 – 1,1 = – 3,9 minutes

$$= (-3 \times 60) + (-0{,}9 \times 60)$$

$$= -180 - 56$$

$$= -236 \text{ sec.arc.}$$

Deviation of star is always subtracted or minus, it mean that the apparent of star is always looks higher than actual star.

Einstein's prediction or according to general relativity, the deviation of Regulus, as the star observed being the closest to the Sun, was about 1,75 sec. arc. But we get results as prediction at 2017 eclipse is about 236 sec.arc. It is more than 100 times greater than Einstein's prediction.

Let's look in Figure 7.3. At the height of eye of 2.5 meters, the DIP correction is about – 2,8 minutes, it's the same with 168 sec.arc. We can see that the DIP correction increase in accordance with the increasing of height of eye. For example, at the height of eye of 18,5 meters, the DIP correction is about -7,6 minutes, it's the same with 456 sec.arc.

In the previous discussion, we know in the F.W.Dyson document that we can't find data of the height of eye or height of the place of observations, neither an observer at Principle Island nor an observer at Sobral. If they really take into account the apparent altitude correction, the result would be more than 100 times greater than Einstein's prediction.

It is important to noted that by using telescope or camera, it is really image of 2-dimensional and apparent position of star, not actual position.

Now, our interest at the moment is to re-examined the result at Madras, Oregon, with the results at Kentucky if 2017 eclipse at the maximum total solar eclipse.

Event	Date	Time (UT)	Alt	Azi
Start of partial eclipse (C1) :	2017/08/21	16:52:44.0	60.3°	145.5°
Start of total eclipse (C2) :	2017/08/21	18:20:29.0	63.9°	191.7°
Maximum eclipse :	2017/08/21	18:21:49.2	63.8°	192.5°
End of total eclipse (C3) :	2017/08/21	18:23:09.2	63.7°	193.2°
End of partial eclipse (C4) :	2017/08/21	19:47:54.1	54.6°	230.4°

Figure 7.4: Maximum eclipse over Kentucky (NASA Eclipse Website).

NASA Eclipse Website informs us:

Altitude of the Sun/Moon

Lat.: 37.5763° N
Long.: 89.1108° W

Total Solar Eclipse
Duration of Totality: 2m40.2s
Magnitude: 1.015
Obscuration: 100.00%

Altitude:

Start of partial eclipse : 60.3°
Start of total eclipse : 63.9°
Maximum eclipse : 63.8°

End of total eclipse : 63.7°
End of partial eclipse : 54.6°

Altitude Start of total eclipse: 63.9°, and altitude End of total eclipse: 63.7°, and at that time during eclipse we see Regulus very close to the Sun/Moon. It is mean that apparent altitude of Regulus is the same with the Sun/Moon during maximum total eclipse: 63.8°.

Now we ready to find deviation of Regulus in nautical almanac, see in Figure 7.5.

ALTITUDE CORRECTION TABLE

Stars and Planets		DIP			
App.Corr.	App.Corr.	Ht	Corr	Ht	Corr.
40.08 −1,1 42.44		2.4 (−2.8) 2.6		Venus 0 +0.1 60	
60.26 (−0.5) 65.06		18.4 −7.6 18.8		Mars 0 +0.1 60	

Figure 7.5

Deviation of Regulus = − 0.5 − 2.8 = − 3.3 minutes.

$$= (3 \times 3) + (0.3 \times 60)$$

$$= 198 \text{ sec.arc.}$$

According to general relativity the deviation of Regulus should be about 1.75 seconds of an arc, but prediction at the maximum 2017 total solar eclipse is about 198 sec.arc, it's mean more than 100 times greater than Einstein's prediction. The results of prediction almost the same with the first prediction during start of eclipse over Madras, Oregon.

1.4. Astronomical Data Proves Spacetime is False.

It was expected from 2017 solar eclipse in USA, that we can see 4 planets: Venus, Jupiter, Mars and Mercury. Actually, this natural phenomena is the best opportunity to prove Einstein's idea on spacetime. The 2017 solar eclipse can be illustrated in Figure 8.1.

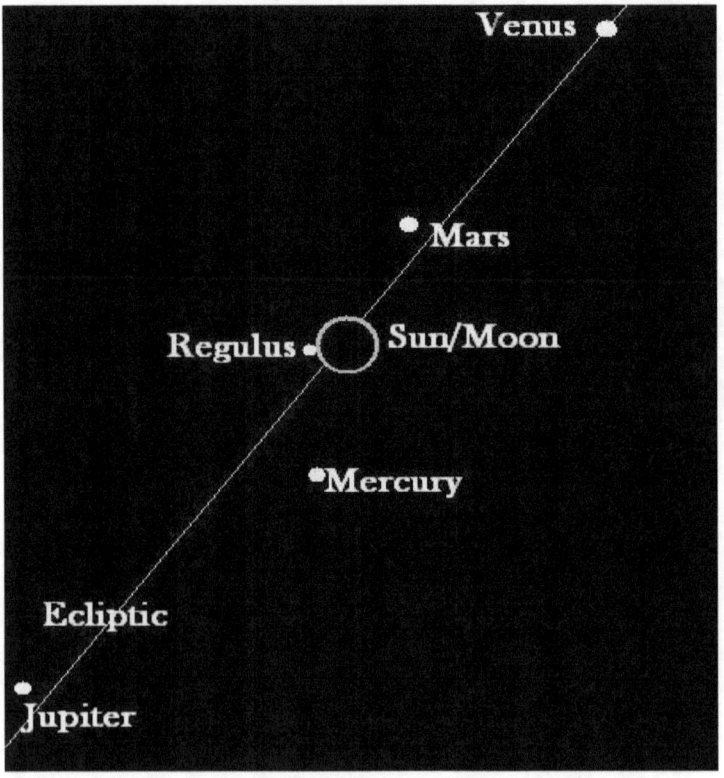

Figure 8.1

In Figure 8.1, we can see from left to right: Jupiter, Regulus, Mercury, Mars, and Venus. We select 4 objects: Mercury, Regulus,

Mars, and Venus, see in Figure 8.2.

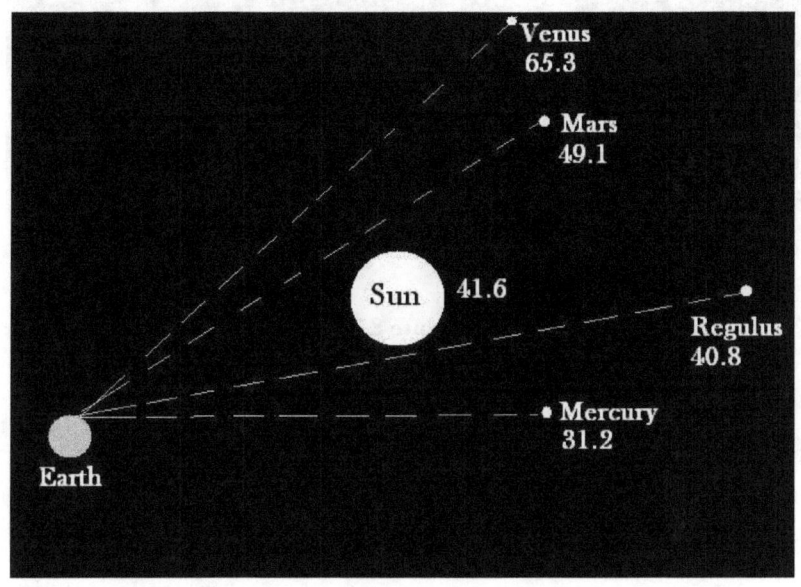

Figure 8.2

From the Figure 8.2 their light tracks can be pictured in the curvature of space, as seen in Figure 8.3.

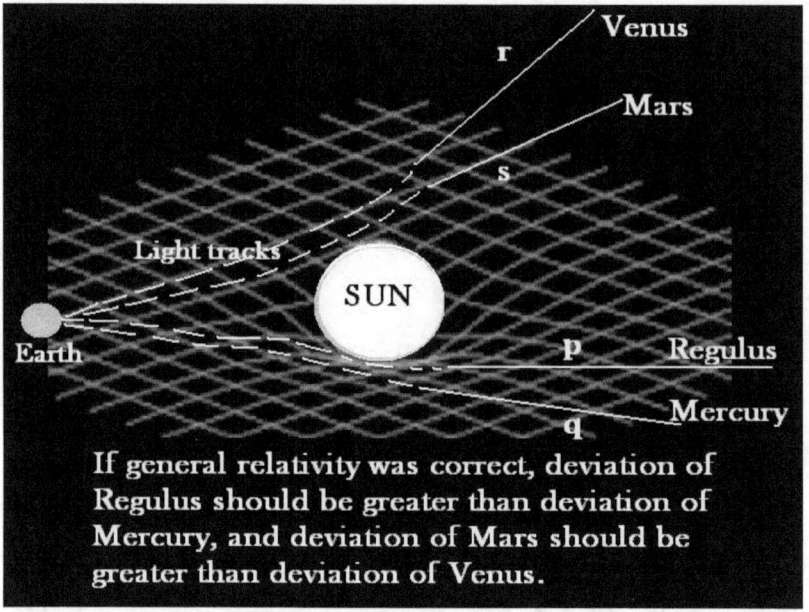

Figure 8.3.

In Figure 8.3, p is the light track of Regulus, q is the light track of Mercury, s is the light track of Mars, and r is the light track of Venus.

Deviation of Regulus: dp.

Deviation of Mercury: dq.

Deviation of Mars: ds.

Deviation of Venus: dr.

According to general relativity, the star light visible around the Sun would be bent inwards, toward the Sun at the time when passing through the gravity field of the Sun. Einstein calculated the level of their deviation and predicted that for the stars observed being the closest to the Sun, their deviation was about 1.75 seconds of an arc.

Therefore, deviation of Regulus (dp) should be about 1.75 sec.arc, and the deviation of Regulus (dp) should be greater than deviation of Mercury, or dp > dq. In the same way the deviation of Mars (ds) should be greater than deviation of Venus (dr), or ds > dr.

Now we can start to validate Einstein's spacetime. To do this, see the astronomical data of apparent altitude of star and planets:

Mercury: 31.2,

Regulus : 40.8,

Mars :   49.1

Venus :   65.3

It is important to note that apparent altitude correction of Mars and Venus can not apply the correction of star. Especially for apparent altitude correction of Venus and Mars, in the Nautical Almanac of 2017 as seen on Figure 8.4..

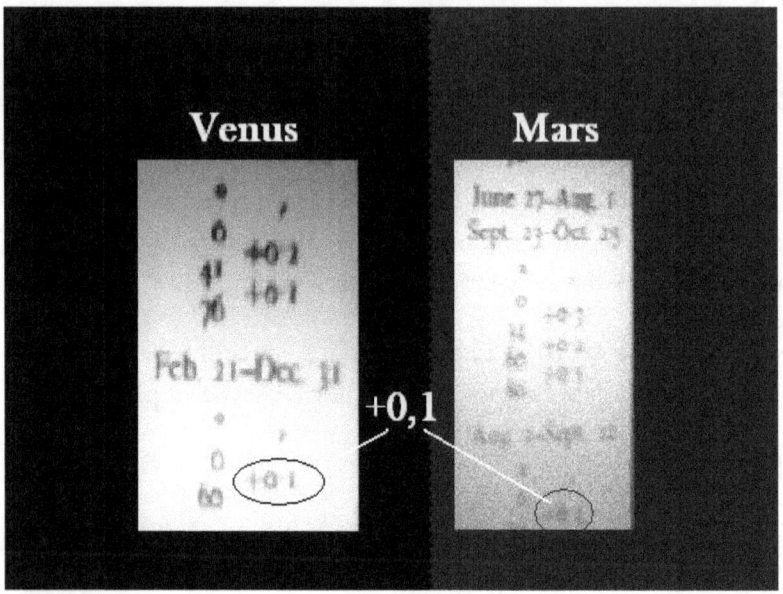

Figure 8.4

The table shows correction for Venus is the same with Mars: + 0,1. Now we can apply height of eye of 2,5 meters, and then calculate the deviation of Mercury, Regulus, Mars, and Venus. The results can be seen in Figure 8.5.

Apparent Altitude Correction

Stars and Planets		DIP	
App.Corr	App.Corr	Ht.Corr	Ht.Corr
30.24 -1.6 32.00	Venus	2,4 -2,8 2,6	
37.48 -1.2 40.08 -1.1 42.44	0 +0.1 60	Mercury: -1.6 - 2.8 = - 4.4 Regulus: -1.1 - 2.8 = - 3.9 Mars: +0.1 -2.8 = - 2.7	
60.28 -0.5 65.08 -0.4 70.11	Mars 0 +0.1 60	Venus: +0.1 -2.8 = - 2.7	

Figure 8.5.

Figure 8.5 shows the results:

1. Deviation of Regulus (dp) is: -3.9 minutes, or -234 sec.arc. It is more than 100 times greater than 1.75 sec.arc as Einstein's prediction. It's proof that general relativity is wrong.

2. dp is -3.9 minutes, while dq is -4.4, or dp < dq. It's mean that light doesn't travel as a curve. It's proof that a curvature space is false.

3. dr is -2.7 and ds is the same -2.7. It's also proof that a curvature space is false.

Thus, a spacetime as a model really do not match up to astronomical data of 2017 elipse. Now we have a valid astronomical data of 2017 solar eclipse, in USA, August 21, that proof spacetime is nonsense.

2. EXPERIMENTS: KNOWING THE RESULT THEY WANTED TO GET.

Gravity Probe B: Mission Impossible?

NASA's Gravity Probe B (GP-B) mission has confirmed two key predictions derived from Albert Einstein's general theory of relativity, which the spacecraft was designed to test. The experiment, launched in 2004, used four ultra-precise gyroscopes to measure the hypothesized geodetic effect, the warping of space and time around a gravitational body, and frame-dragging, the amount a spinning object pulls space and time with it as it rotates. GP-B determined both effects with unprecedented precision by pointing at a single star, IM Pegasi, while in a polar orbit around Earth. [1]

Figure 1.1: Artist concept of Gravity Probe B spacecraft in orbit around the Earth. Image Credit: Stanford

According to NASA, the Gravity Probe B gyroscopes are the most perfect spheres ever made by humans. If these ping pong-sized balls of fused quartz and silicon were the size of the Earth, the elevation of the entire surface would vary by no more than 12 feet.

Gravity Probe B (GP-B) is a NASA physics mission to experimentally investigate Albert Einstein's 1916 general theory of relativity—his theory of gravity. GB-B uses four spherical gyroscopes and a telescope, housed in a satellite orbiting 642 km (400 mi) above the Earth, to measure in a new way, and with unprecedented accuracy, two extraordinary effects predicted by the general theory of relativity (the second having never before been directly measured):

1. The geodetic effect—the amount by which the Earth warps the local spacetime in which it resides.

2. The frame-dragging effect—the amount by which the rotating Earth drags its local spacetime around with it.

In this discussion, my interest is finding the relationships between spacetime and atmospheric medium, as I know that in general relativity Einstein had ignored the atmospheric medium around a massive body in the sky.

Earth's gravity pulls all the objects in atmosphere toward the Earth, and all the objects follow the rotation of the Earth. In the same way the satellite Gravity Probe B has orbit at the altitude 400 miles (642 km) above the Earth, remains in orbit at thermosphere/exosphere, and doesn't escape to space, it's because Earth's gravitational force. That is a fact, no doubt about it.

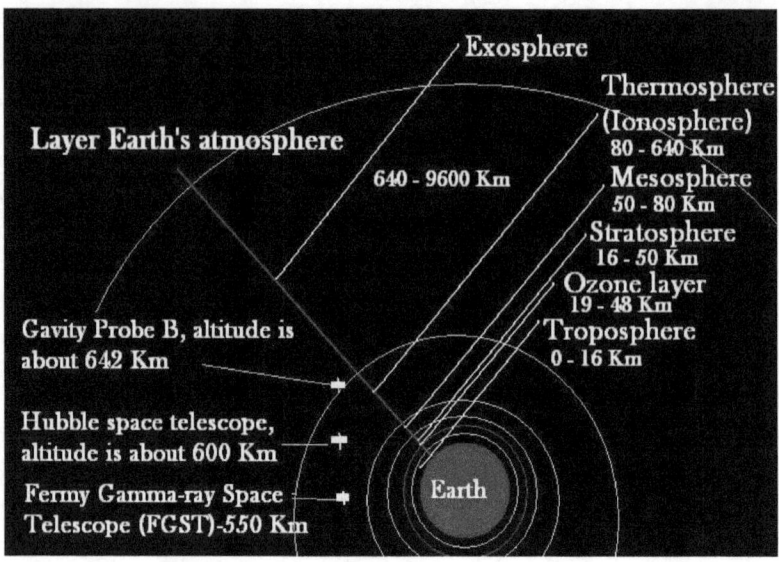

Figure 1.2

Something logical fallacies if they ignored the effects of Earth's gravitational force, and then, they try to prove 'gravity is nothing about force' according to Einstein's general theory of relativity, in the case of the geodetic effect and the frame-dragging effect.

The above figure shows orbit of Gravity Probe B in the Earth's atmosphere. Of course, the Earth's atmosphere is not spacetime. Thus, in this case, it can be seen as a false equivalence—describing a situation of logical and apparent equivalence, when in fact there is none.

"In 2011 a spacecraft known as Gravity Probe B successfully observed this effect due to the Earth.

Despite the challenges, Gravity Probe B confirmed the Earth's gravitational curvature of space to within 1% of predictions, and confirmed frame dragging to within 19%."(Brian Koberlein,)[2]

How could they detect the geodetic effect and the frame-dragging effect in the Earth's atmosphere, although the Gravity Probe B gyroscopes are the most perfect spheres ever made by humans?

If they think spacetime is the same with atmospheric medium, so they should change Einstein Field Equation of gravitation (EFE). As we know, the meaning of EFE is :

Curvature of spacetime = density and flux of energy and momentum.

This model is highly speculative and can't answer the question: "How is it possible for a mass to curve spacetime? So, they can change this model become:

Atmospheric medium = density and flux of energy and momentum.

Mile Mathis wrote in his article entitle GRAVITY PROBE B and space-time

"In a nutshell, what the Gravity Probe experiment did is measure the tilt of little gyroscopes.

If the tilt is zero, no curvature of space-time. If the tilt is not zero, we are supposed to have proof of curvature. The gyroscope tilts because space is curved.

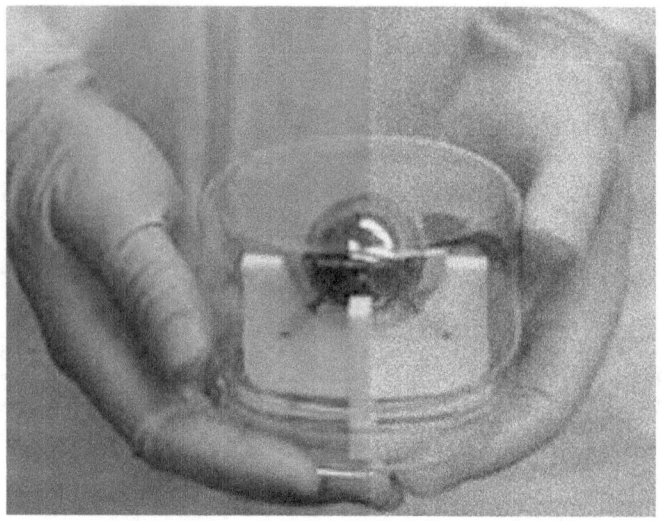

Figure 1.3

The primary problem is that there is absolutely no effort in this experiment to consider, mention, or try to block the main cause of that tilt. It is simply assumed that any non-zero outcome is proof positive of their theory and that any tilt that does not match their needed numbers is only an anomaly or "observation" that can be explained away later. That is horrible science, no matter how you look at it.

Gravity Probe B project that was given a grade of F in a NASA review in 2008 by a group of senior advisors, and denied any further funding since, "the reduction in noise needed to test rigorously for a deviation from general relativity is so large that any effort ultimately detected by this experiment will have to overcome considerable (and in our opinion, well justified) scepticism in the scientific community". They continue:

"The noisy data meant that GPB could not measure the effects as precisely as astronomers had by firing laser beams at mirrors left of the Moon by the Apollo astronauts."

Measurement Using VLBI.

RADIO observations using very-long-baseline interferometry (VLBI) can measure the deflection of electromagnetic radiation by the Sun's gravitational field with an accuracy of better than 1 milliarcsecond, and can thus be used to test General Relativity. [3]

Figure 2.1

Between the two theories of gravitation, Newton's Law of Gravitation and Einstein's theory of gravity (general theory of relativity) both of theories can not both be true. Both can be wrong or only one can be right.

In this regard, I was interested to know some of the opinions of physicists, for example from the article in Nature: **Arthur Eddington was innocent!**

"One of the more recent victims of this revised thinking is the 'confirmation' of Einstein's theory of general relativity, offered in

1919 by the British astronomer Arthur Eddington. Eddington reported observing the bending of light during a total eclipse, as predicted by Einstein. But some have claimed that he cooked his books to make sure that Einstein was vindicated over Newton, because Eddington had already decided that this must be so.

Now, even physicists who celebrate Einstein's theory commonly charge Eddington with over-interpreting his data.

In his Brief History of Time, Stephen Hawking says of the result that: "Their measurement had been sheer luck, or a case of knowing the result they wanted to get." Hawking reports the widespread view that the errors in the data were as big as the effect they were meant to probe. Some go further, saying that Eddington deliberately excluded data that didn't agree with Einstein's prediction." [4]

And in the medium, we read the opinion of Professor Brian Koberlain in the article: **The Strangest Theory We Know Is True**, After 99 years, Einstein's greatest scientific achievement is undefeated.

"Using the equivalence principle, Einstein developed a gravitational theory even more strange than special relativity. In his theory the fabric of space and time can be bent and twisted by the presence and motion of masses. One of the first tests of the gravitational curvature of space was the deflection of starlight during a solar eclipse, first observed by Edenton in 1919. Eddington's results supported Einstein's model, but not very strongly. Given the radical approach of general relativity, Eddington results were initially disputed by some. But subsequent observations confirmed Einstein's predictions."(Prof.Brian Koberlain)

Somewhat surprising is the opinion of Sabine Hossenfelder, which is clearly blames Arthur Eddington.

"As history has it, Eddington original data actually wasn't good enough to make that claim with certainty. His measurements had huge error bars due to bad weather and he also might have cherry-picked his data because he liked Einstein's theory a little too much. Shame on him." (Sabine Hossenfelder) [5]

But finally most of them accepted Einstein's theory of gravity, without rejected Newton's theory of gravity. Sabine Hossenfelder tells us the reasons:

"By the 1990s, one didn't have to wait for solar eclipses any more. Data from radio sources, such as distant pulsars, measured by very long baseline interferometer (VLBI) could now be analyzed for the effect of light deflection. In VLBI, one measures the time delay by which wave fronts from radio sources arrive at distant detectors that might be distributed all over the globe. The long baseline together with a very exact timing of the signal's arrival allows one to then pinpoint very precisely where the object is located—or seems to be located. In 1991, Robertson, Carter & Dillinger confirmed to high accuracy the light deflection predicted by General Relativity by analyzing data from VLBI accumulated over 10 years."

Einstein online website: "With repetitions of eclipse measurements over the next half century, astronomers were able to improve on the accuracy of these first results by only about a factor two, yielding a confirmation of general relativity to within about ten percent. The breakthrough came in 1967 with the realization that simultaneous measurements with a set of radio telescopes (especially, "Very Long Baseline Interferometry") could be used to measure light deflection with much greater accuracy."(einstein-online.info) [6]

At the same time, and there are two conflicting theories, but both theories are considered to be true. Of course, it is something unusual happens in science.

How could it be happened in science?

Here my interest in this writing to finding the answer, and may be I've found the answer. What are the reasons that caused it happen, it is because the VLBI measurements can provide the data as if someone wants to get the result.

Wikipedia informs us some of the scientific results derived from VLBI include:

1.High resolution radio imaging of cosmic radio sources.

2.Imaging the surfaces of nearby stars at radio wavelengths (see also interferometry)—similar techniques have also been used to make infrared and optical images of stellar surfaces.

3.Definition of the celestial reference frame.

4.Motion of the Earth's tectonic plates.

5.Regional deformation and local uplift or subsidence.

6.Variations in the Earth's orientation and length of day.

7.Maintenance of the terrestrial reference frame

8.Measurement of gravitational forces of the Sun and Moon on the Earth and the deep structure of the Earth

9. Improvement of atmospheric models

10. Measurement of the fundamental speed of gravity

11. The tracking of the Huygens probe as it passed through Titan's atmosphere, allowing wind velocity measurements.

Let's note point 1 and point 8. At point 1 High resolution radio imaging of cosmic radio sources, can be used to test general relativity as Letters to Nature-Nature 349, 768–770 (28 February 1991). Its mean Einstein's gravity 'nothing about force' is true.

At point 8 Measurement of gravitational forces of the Sun and Moon on the Earth and the deep structure of the Earth. Its mean Newton Law of Gravitation is correct.

In this case, they can prove that both theories are true. But it can not happens in science. For example, if someone says:"VLBI confirmed to high accuracy both Newton's gravity and Einstein's gravity", this is something logical fallacies. Both theories can be wrong, but only one can be right.

Primary problem of general relativity

Actually, the primary problem of general relativity is not test of hypothesis, but the hypothesis itself. As widely known, Einstein hypothesis of general relativity is called 'deflection of light by the Sun'. But, what's the meaning of deflection of light (starlight) by the Sun? Let's look at in Figure 2.1, 2.2, and 2.3 below.

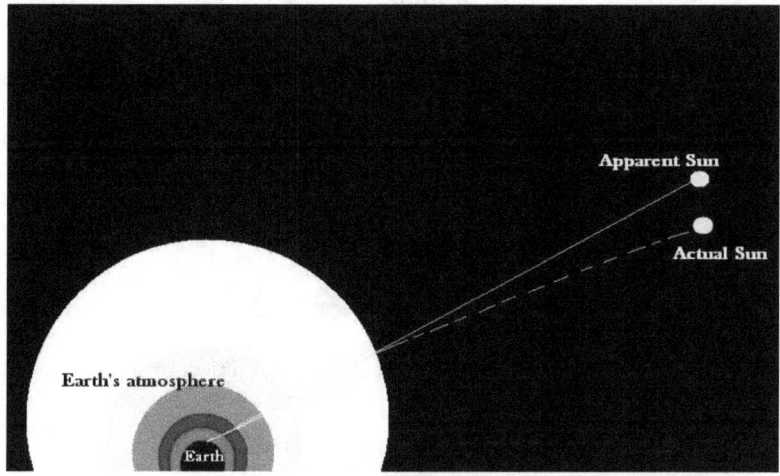

Figure 2.1: Deflection of Sunlight by Earth's atmosphere.

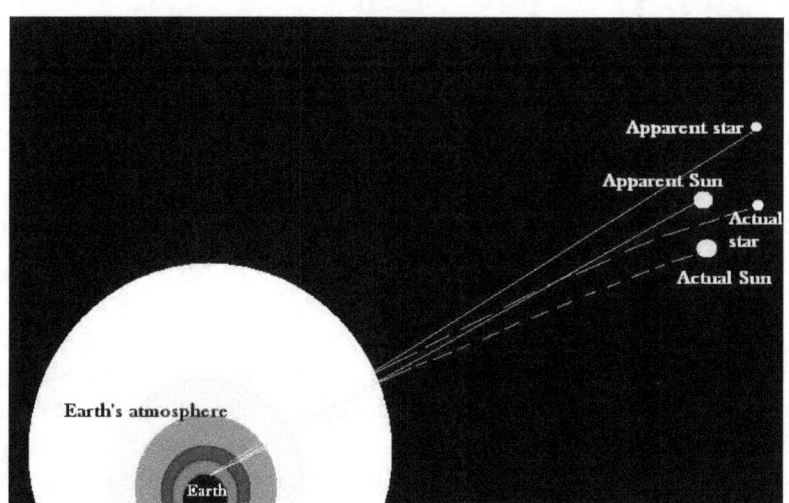

Figure 2.2: Deflection of Sunlight and Starlight by Earth's atmosphere.

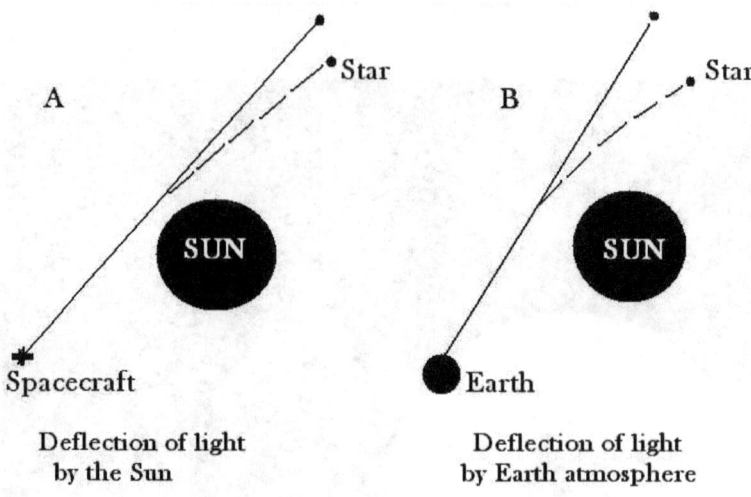

Figure 2.3: Defection of Starlight

The three illustrations above give a simple explanation about the deflection of light: Sunlight and Starlight. In Figure 2.3, A is deflection of starlight by the Sun as seen from space by space telescope mounted on the spacecraft. B is deflection of starlight as seen from the Earth. It is not deflection of light by the Sun, but deflection of light by Earth's atmosphere.

What then? The three illustrations above could explain as a simple explanation about invalidity of Einstein hypothesis on his idea of gravity (general relativity).

If someone says:"VLBI confirmed to high accuracy both Newton's gravity and Einstein's gravity", this is something logical fallacies. Both theories can be wrong, but only one can be right. If someone says:"Einstein was right", must be able to says: "Newton was wrong" and shows evidence. If not, that is useless.

Sextant is the most accurate telescope.

Need to know the basic level of astronomy, in order to know that collecting data of stars, and then comparing data made at another time; can not be use to measure deviation of starlight. Measuring the angle in astronomy applies direct measuring and instantaneous. It doesn't matter using a sophisticated software, if doesn't meet requirements in the basic principle of scientific method in the field of astronomy; should be classified as a non-scientific.

Thank to Sir Isaac Newton

The idea of "a small telescope" or a sextant employing a movable mirror was first conceived by Isaac Newton in 1700. Workable instruments were made in 1730; sextant in this form has been in use for 250 years, and will be used into the foreseeable future. The name is derived from the Latin sextant, or sixth part of a circle. Due to the arrangement of the optic, the sextant will actually measure angles up to one third of a circle, or 120 degrees. The octants and quadrant are similar instruments, with ranges of 90 and 180 degrees.

A sextant can also be used to measure the lunar distance between the moon and another celestial object (such as a star or planet) in order to determine Greenwich Mean Time and hence longitude. The principle of the instrument was first implemented around 1730 by John Hadley (1682–1744) and Thomas Godfrey (1704–1749), but it was also found later in the unpublished writings of Isaac Newton (1643–1727)-(wikipedia.org) [7]

On the first glance you might think that a sextant looks pretty complicated, but it really isn't. There are only three basic parts, as shown in Figure 3.1 below, and the parts of the "small telescope" are clearly described on wikipedia.

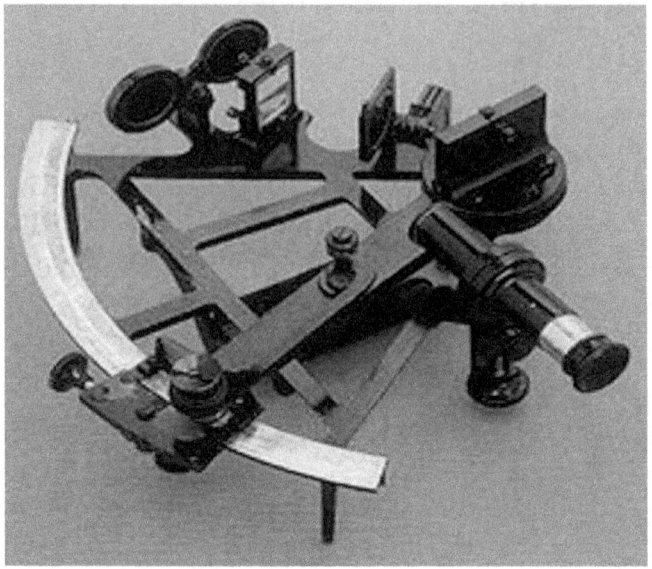

Figure 3.1: A Sextant

Radio Telescope

What are Radio Telescopes? We use radio telescopes to study naturally occurring radio light from stars, galaxies, black holes, and other astronomical objects. We can also use them to transmit and reflect radio light off of planetary bodies in our solar system. These specially-designed telescopes observe the longest wavelengths of light, ranging from 1 millimeter to over 10 meters long. For comparison, visible light waves are only a few hundred nanometers long, and a nanometer is only 1/10,000th the thickness of a piece of paper!

Naturally-occurring radio waves are extremely weak by the time they reach us from space. A cell phone signal is a billion billion times more powerful than the cosmic waves our telescopes detect.(public.radio-telescopes) [8]

Figure 3.2: Flying southward over the Very Large Array in central New Mexico shows the flatness of the Plains of San Agustin where the array is sited. At over 7,000-feet elevation, this ancient lakebed provides the space needed to lay 40 miles of double-track rails required to extend the VLA's Y-shape to give us its highest resolution capability.

The farther we separate our radio antennas, the larger the telescope they mimic. The phase shifts they see are even greater, which means their narrower overlap is a finer detail view of the sky. With this level of accuracy, radio telescopes spread very far apart can pinpoint exact locations of radio objects in space, including distances from Earth. We call this system Very Long Baseline Interferometry, or VLBI for short. The Very Long Baseline Array (VLBA) is the world's largest VLBI system dedicated to full-time research.

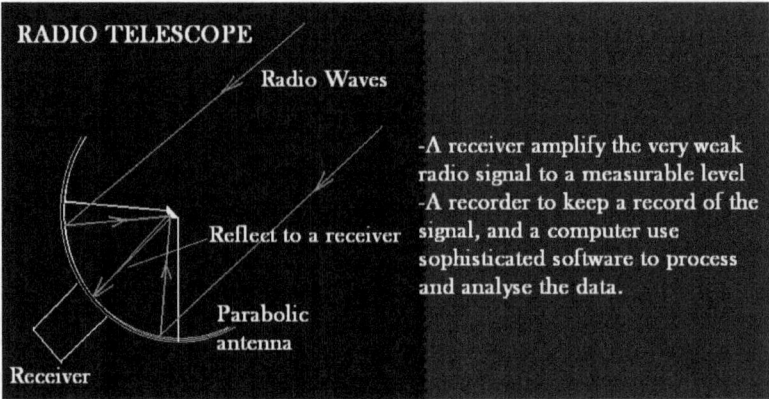

Figure 3.3

From the above discussion and knowing how the radio telescope works, we can imagine the difficulty to measure the deviation of starlight using radio telescope. It's about the precession of angle, not just an image of the objects.

The main problem is the weak incoming signal, and the other difficulty the observations can not be done just once, but many times, and maybe 1000 times of observations. Obviously, needs a long time.

We'll probably never really know the computer sophisticated software of VLBI. As we know, it is very difficult to improve the resolution of the telescope only by increasing the size of the incoming radio waves; disturbances in the atmosphere limit the resolution of radio telescope. Radio telescope should be taking into account the effect of refraction and aberration of light (radio waves) depending on elevation of location of the radio telescope.

"According to Will, an analysis in 2004 of over 2 million VBLI observations has shown that the ratio of the actual observed deflections to the deflections predicted by general relativity is 0.99992 ± 0.00023. Thus the dramatic announcement of 1919 has been retro-actively justified." [9]

This website informs about the difficulty of performing precise measurements of optical starlight deflection during an eclipse; can be gathered from the following list of results:

Optical Deflection of Starlight During Eclipses

Date	Location	arc secs
29 May 1919	Sobral	1.98 ± 0.16
	Principe	1.16 ± 0.40
21 Sep 1922	Australia	1.77 ± 0.40
		1.42 to 2.16
		1.72 ± 0.15
		1.82 ± 0.20
9 May 1929	Sumatra	2.24 ± 0.10
19 June 1936	USSR	2.73 ± 0.31
	Japan	1.28 to 2.13
20 May 1947	Brazil	2.01 ± 0.27
25 Feb 1952	Sudan	1.70 ± 0.10
30 Jun 1973	Mauritania	1.66 ± 0.19

Over two million VBLI observations! Of course, there is lot of angle of incidence the incoming radio waves. This is very surprising and amazing. *But, is this 'a fair game' in the sense that no other people could easily test the accuracy of the results, using the same tools?*

This claims actually useless. That is something logical fallacy: argumentum ad lap idem-dismissing a claim as absurd without demonstrating proof for its absurdity.

Last but not least, a lot of details about the bending of light, but does not explain the bending of light as seen from space or from the Earth:

Deflection and Delay of Light. Everybody knows that light travels in straight lines, but while that is its natural tendency light can be deflected by lenses, mirrors, and by gravitational fields. Newtonian mechanics predicts that a particle traveling at the speed of light which just grazes the edge of the Sun will be deflected by 0.875 seconds of arc. That means that the image we see of a star will be displaced away from the Sun by this angle. The figure below shows this with the black showing the situation when the Sun is not close to the star. When the Sun is nearly blocking the star its image is deflected outward giving the red image. This Newtonian model also predicts that the gravitational attraction of the Sun will make light travel faster close to the Sun, so according to Newton the deflected light arrives before the undeflected light. The figure shows the red light pulse arriving before the black light pulse. Of course the travel time for starlight is very hard to measure, and the deflection of starlight can only be measured during a total eclipse of the Sun. The deflection angle is actually very small, and in the figure it has been increased by a factor of nearly 10,000 for clarity.(www.astro.ucla.edu)

From the above website we can read a statement: General Relativity Wins Again. 17 Sep 2009—Today's Nature has a letter explaining the anomalous precession of the orbit of DI Herculis by Albrecht et al. 2009, Nature, 461, 373. A preprint is also available. It turns out that the spin axes of the stars are quite mis-aligned with the orbit, leading to tidal torques that explain why the precession was slower than the prediction of General Relativity.

'General Relativity Wins Again', this is also something logical fallacy : argumentum ad lap idem-dismissing a claim as absurd without demonstrating proof for its absurdity (This claim doesn't demonstrated Newton's gravity is wrong).

Prediction of General Relativity Doesn't Work.

Why Einstein will never be wrong. One of the benefits of being an astrophysicist is your weekly email from someone who claims to have "proven Einstein wrong". These either contain no mathematical equations and use phrases such as "it is obvious that..", or they are page after page of complex equations with dozens of scientific terms used in non-traditional ways. They all get deleted pretty quickly, not because astrophysicists are too indoctrinated in established theories, but because none of them acknowledge how theories get replaced.(Prof.Brian Koberlein). [10]

Of course not, obviously not because astrophysicists are too indoctrinated in established theories, but because astrophycists (and physicists) has no experiences in celestial navigation as a navigator at sea (not less than two years).

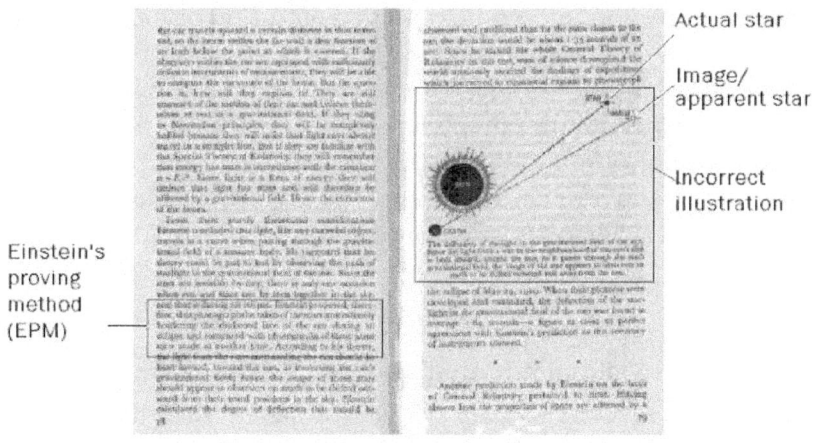

Public domain image - Archive.Org

Figure 4.1: Invalidity hypothesis; EPM isn't scientifically correct and illustration incorrect

Two Fatal Mistakes: *He wants measuring deflection of light by the Sun;*

but he proposed test measuring deflection of light by Earth's atmosphere; he had not realized about that. Ironically, this test is not scientifically correct and deeply wrong. And ....London News, November 22, 1919 had published incorrect illustration around the world, that is make a lot of authors on the popular scientific like Lincoln Barnett got wrong in his book.

Modern scientists are doing too much trusting and not enough verifying, that's why many of scientists was involved to get wrong, for example, this explanation is incorrect:

Equatorial Coordinate System. This is the preferred coordinate system to pinpoint objects on the celestial sphere. Unlike the horizontal coordinate system, *equatorial coordinates are independent of the observer's location and the time of the observation.* This means that only one set of coordinates is required for each object, and that these same coordinates can be used by observers in different locations and at different times.(Cosmos) [11]

Equatorial coordinates are not independent of the observer's location and the time of the observation, and that these same coordinates can not be used by observers in different locations and at different times. Each place in the Earth lets say Oxford in UK and Principe Island in West-Africa has it own equatorial coordinates, these coordinates depend on geographic position: i.e. latitude and longitude. The same of object in the sky, a star, that can be can be photographed from Oxford and Principe Island in the different time; can not be compared for calculate deviation of the star.

Alphonsus Kelly, an Ireland engineer, in his lecture at Trinity College, Dublin, on February 15, 1996, stated that the Einstein's Relativity theory might be wrong. Kelly revealed the experiment of Sagnac, the French physicist in the year 1914, showing that the time taken by a light to complete one rotation is found to be different from the time taken by one rotation in opposite direction. The Sagnac's experiment proved that the speed of light was not constant. It is different from Einstein's theory stating that the light velocity is constant.

Reaction was given by an English astronomer, Arnold Wolfendale, stating that to design an experiment to prove that the light velocity is not constant is really a difficult thing to do.

" You cannot demolish a very strong theory such as Einstein's Relativity only be based on a cheap experiment," said Wolfendale referring to Kelly's opinion indicating that Einstein was wrong just based on Sagnac's experiment. Wolfendale added that the particle accelerator all over the world have proven the truth of Einstein's Relativity.

"We are engineer, will never give up, "said Kelly to The Times. " I know there is a priest preaching a mystery not known to him, and I think the physicists do the same thing."

In the year of 2015 scientists show experimental evidence as Prof. Howard Wiseman said Einstein was WRONG.

Prof. Howard Wiseman: Einstein was WRONG.

'Spooky' quantum experiment shows that the measurement of a photon affects its location. Previous experiments have shown entanglement with two particles. But this is the first to show the entanglement of a photon with itself. The study reveals how when a light photon is observed it changes state

Einstein didn't believe this could happen as it violates theory of relativity. Albert Einstein may have been a genius, but even he could get it wrong sometimes.

In the 1920s and 1930s, Einstein said he couldn't back the strange theory that the measurement of a particle actually affects its location. Now a team of scientists from Japan and Australia have proven that this 'spooky action at a distance' takes place in a photon. [12]

Why Einstein will never be wrong

Stop for a moment here, because it relates with an interesting article entitled Why Einstein will never be wrong, on Universe Today Website. This article is very interesting and there are some points that need to be discussed.

"One of the benefits of being an astrophysicist is your weekly email from someone who claims to have "proven Einstein wrong". These either contain no mathematical equations and use phrases such as "it is obvious that..", or they are page after page of complex equations with dozens of scientific terms used in non-traditional ways. They all get deleted pretty quickly, not because astrophysicists are too indoctrinated in established theories, but because none of them acknowledge how theories get replaced."(Prof.Brian Koberlein)

In this article there are four points that are interesting:

Statement 1." ....but because none of them acknowledge how theories get replaced."

Statement 2:"Einstein's gravity will never be proven wrong by a theory."

Statement 3: "The other way to trump Einstein would be to develop a theory that clearly shows how Einstein's theory is an approximation of your new theory, or how the experimental tests general relativity has passed are also passed by your theory. Ideally, your new theory will also make new predictions that can be tested in a reasonable way."

Statement 1 explains if someone had found evidence Einstein's theory is wrong; should be conveying a new theory. But statement 2 said "Einstein's gravity will never be proven wrong by a theory." It means the author negates his first statement itself.

I'm so interest in statement 3; advocated a new theory that is consistent with the theory of Einstein (how Einstein's theory is an approximation of your new theory).

Here a question: How does if someone show invalidity of Einstein's hypothesis; Einstein's hypothesis can not be tested in a reasonable way, or to be more precisely, Einstein's hypothesis does not meet requirement of the scientific method?

Actually, answers to this question had been anticipated by Einstein himself. Einstein courageously said "The chief attraction of the theory lies in its logical completeness. If a single one of the conclusions drawn from it proves wrong, it must be given up; to modify it without destroying the whole structure seems to be impossible." [13].

So far, this writing has not answered or explained title Experimental Evidence Predictions of General Relativity Do not Work. This title in connection with Statement 4 of Brian Koberlein's writing:

"It will be proven wrong by experimental evidence showing that the predictions of general relativity don't work. So unless you have experimental evidence that clearly contradicts general relativity, claims of "disproving Einstein" will fall on deaf ears".

Experimental Evidence

Let us note, if general theory of relativity was correct, then the light from stars that passed closest to the sun would show the greatest degree of bending, and the stars whose light tracks are very far from the sun have their lights not being bent. The stars whose lights are not being bent means that there is no difference between the apparent position and the true position of the stars.

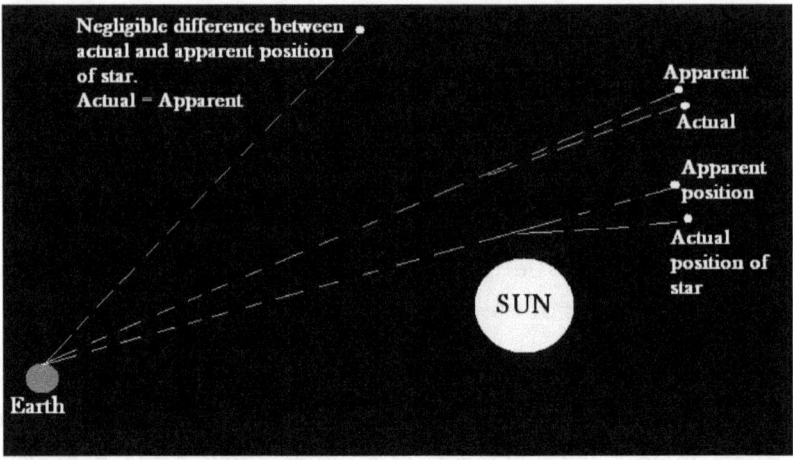

Figure 4.2: If Einstein's theory of relativity was correct, then the light from stars that passed closest to the sun would show the greatest degree of "bending." – In UndsciBerkeleyEdu admin had not realized illustration also incorrect. Apparent position should be look higher than actual position. [14]

If Einstein's theory of relativity was correct, then the light from stars that passed closest to the sun would show the greatest degree of "bending" and the stars whose light tracks are very far from the sun "negligible difference between actual and apparent positions of star"- See Figure 4.2.

Do not just look at what it is written, let us see what is not written; the consequences of general relativity is that, if being consistent with this theory, it means that all stars visible at night time are at the appearance of the stars on their true positions, because the said stars do not pass through the field of gravity.

It means that general relativity predicted or determined that all stars visible at night time are at the appearance of the stars on their true / actual positions; in other words, general relativity predicts the occurrence of 'negligible difference between actual and apparent positions of star'

This is certainly incorrect if it is seen from the astronomical scientific point of view. The stars in the sky at night time and seen by the observers, all are stars on apparent positions, not on their true / actual positions.

There is millions of experimental evidence about it.

1. The navigator of a vessel at position 50 degrees 52' (minutes) N, 22 degrees 07' (minutes) W takes sight of Acturus, at 1973 June 18d 22h 57m 04s. The sextant reading is 56 degrees 11' (minutes). Index error is 06 minutes, height of eye is 14 feet. The resulting of correction to obtain apparent altitude of star can be found from Table 1 Nautical Almanac of the year 1973:

Taking Sight of Acturus

Star	Sextant Altitude	Zone Time
Acturus	$56^0 11'$	
Index Corr	+06'	1973 June 18d
DIP	-04'	22h 57m 04s
App. Alt	$56^0 13'$	
App. Alt. Corr	-01'	
True Alt	$56^0 12'$	

Figure 4.3: Taking sight of Acturus.

In the Figure 4.3 shows measuring of altitude of star as an experiment. This experiment shows that all of stars visible at night time are on their apparent positions, not on their actual positions. It is because all the calculation to find deviation starlight should be considering the effect of light's refraction: i.e. astronomical refraction and terrestrial refraction.

Since the 18th century, astronomers had published the Nautical Almanac; can be used to facilitate the calculation. Nautical Almanac provides some of correction table, including correction for apparent altitude of Sun, Stars, and Planets (astronomical refraction), and height of eye an observer. (terrestrial refraction).This is recognized by the entire navigator in the world for a long time ago.

2. What's a mean? It means that general relativity predicted or determined that all stars visible at night time are at the appearance of the stars on their true / actual positions; IS FALSE. The deeper meaning, general theory of relativity is wrong.

In this case someone can say; predictions of general relativity about 'negligible difference between actual and apparent positions of star' don't work.

Predictions of general relativity really doesn't work:

Taking sight of Kochab and Spica

Star	Sextant Altitude		Zone Time
Kochab	47° 19.1'		20-07-43
Spica		32° 34.8'	20-11-26
Index Corr	+2.1'	+2.1'	Taking sight on
Dip Corr	-6.7'	-6.7'	May 16, 1995
(height 48 ft)			
App.Alt	47° 14.5'	32° 30.2'	
Alt Corr/Refc	-0,9'	-1.5'	
True Alt	47° 13.6'	32° 28.7'	

Figure 4.4: Taking sight of Kochab and Spica

Astronomical Data Verify Einstein's Prediction Doesn't Work.

Hans C. Ohanian's Einstein's Mistakes: The Human Failings of Genius:

Almost all of Einstein's seminal works contain mistakes. Sometimes small mistakes—mere lapses of attention—sometimes fundamental failures to understand the subtleties of his own creations, and sometimes fatal mistakes that undermined the logic of his arguments.

The book was reviewed positively in a recent Wall Street Journal article.

A theoretical physicist by training, Mr. Ohanian doesn't write like one. He recounts his chronicle of errors in clear and engaging prose, giving us in the process a short course in the history of modern physics and a witty and provocative account of his subject's life. Anyone who has read the recent biographies of Einstein by Walter Isaacson or Jürgen Neffe may find some of the material familiar, but on the whole "Einstein's Mistakes" is original and fresh. Nor is Mr. Ohanian one of those petty biographers who delight only in turning up the failings—or turning out the dirty laundry—of great men. Rather he notes Einstein's errors for a purpose, showing us why his achievement was all the greater for them. [15]

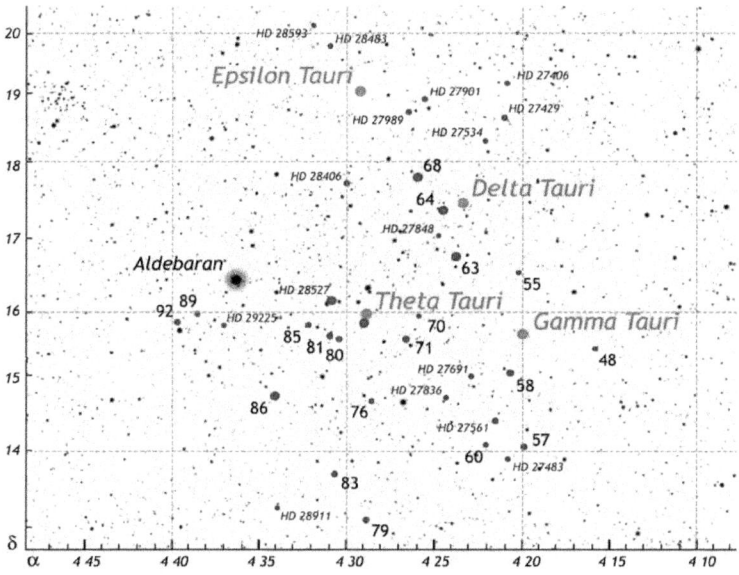

Figure 5.1: Hyade cluster.

Invalidity hypothesis of general relativity

Einstein's Special and General Theory of relativity is widely known as theories in modern physics, as they said, that are pass of every tests for more than 100 years. Stephen Hawking said: "General Relativity was a major intellectual revolution that has transformed the way we think about the universe. It is a theory not only of curved space, but of curved or warped time as well."

I've found there are at least 5 logical fallacies of Einstein's theory of relativity. Of course, some invalid arguments of Einstein as the founder of theories make a great impact to the validity of the two theories: Special and General Theory of Relativity. For example, the experimental test of general relativity called deflection of light by the Sun.

That terms deflection of light by the Sun meaning the deflection as seen from space:

"From these purely theoretical considerations Einstein concluded that light, like any material object, travels in a curve when passing through the gravitational field of a massive body" (Lincoln Barnett, The Universe and Dr.Einstein, London, 1949, Foreword by Albert Einstein himself, page 78).

Actually, that is different between the deflection of light by the Sun as seen from space; and deflection of light by the Sun as seen from the Earth.

"He suggested that his theory could be put to test by observing the path of starlight in the gravitational field of the Sun. Since the stars are invisible by day, there is only one occasion when Sun and stars can be seen together in the sky, and that is during an eclipse. Einstein proposed therefore, that photographs be taken of the stars immediately bordering the darkened face of the sun during an eclipse and compared with photographs of those same stars made at another time."

If the deflection of light by the Sun as seen from the Earth; of course, the effects of Earth's atmosphere can not be ignored. A worthwhile meaning should be noted here; if as seen from the Earth, it is not deflection of light by the Sun but deflection of light by the Earth's atmosphere.

"According to his theory, the light from the stars surrounding the Sun should be bent inward, toward the Sun, in traversing the Sun's gravitational field; hence the images of these stars should appear to observer on earth to be shifted outward from their usual positions in the sky."

From the discussion above is clearly noticeable two fatal mistakes of Einstein; he wants measuring deflection of light by the Sun; but he proposed test measuring deflection of light by Earth's atmosphere; he had not realized about that. Ironically, this test is not scientifically correct and deeply wrong.

In facts, the eclipse experiment was conducted seven times; 1919 eclipse experiment was error, but repeated in the year of 1922, 1929, 1936, 1947, and then in the year of 1952, and again in the year of 1973. All the eclipse experiments with results Einstein's general relativity was right. **Here a big question**: *Repeating the wrong method for decades, how could it be happen in modern science?*

Astronomical data of eclipse in the year 1919–1973

"In theory Einstein's equations allow you to work out exactly how massive objects, such as planets, stars, galaxies, or even black holes affect the spacetime they sit in. In practice though, things aren't quite as straight-forward. Einstein's equations are incredibly difficult to solve—supercomputers are needed to find solutions and coming up with new solutions is an active field of theoretical physics. One of the big current challenges is to figure out what happens to space-time when two very heavy objects, like black holes, collide.

How do we know that Einstein's theory is correct? In the hundred years since its publication, the theory has passed every test it has been subjected to." (David Tong) [16]

Is it true the theory has passed every test it has been subjected to?

If David Tong is correct in his statement, that means the 1921 Nobel Committee is incorrect in his statement:

"without taking into account the value that will be accorded your relativity and gravitation theories after these are confirmed in the future"

This meant that the 1919 eclipse experiment was deemed unable to confirm as announced by Arthur Eddington through London News, November 22, 1919.

Now let's see what exactly happened with the 1919 eclipse experiment. The test was done according to Einstein's proposed test methods, the result will always be wrong; because Einstein ignores the effects of refraction of light: i.e. astronomical refraction and terrestrial refraction.

Einstein calculated the degree of deflection that should be observed and predicted that for the stars closest to the Sun the deviation would be about 1.75 sec. arc. **The important things to be noted**, *the amount of 1.75 sec.arc without mentioning the altitude of the Sun as the object of observation. This is a fatal mistake; or something like a joke; because the deviation of starlight will always vary depending on the altitude of the object of observation from the sea level.*

Moreover, the amount of deviation is very small, and the ability of the telescope at that time, will be very difficult to determine the value of less than 2 sec.arc.

In addition, since the 18th century astronomers published the Nautical Almanac that can be used to facilitate calculations related to starlight deviation. That's really hard to understand they ignore Nautical Almanac of 1919.

Astronomical data of 1919 solar eclipse

We'll easily able to see in the Nautical Almanac of 1919; the deviation of a certain star that closest to the Sun during maximum solar eclipse. Apparent altitude of maximum eclipse is about 70,6 degrees, the astronomical refraction is about 21 sec.arc, it more than 10 times greater than Einstein's prediction.

1919 Eclipse - West Africa
Lat : $2.1089°$ N
Long: $16.875°$ W

Event	Date	Time (UT)	Alt	Azi
Start of partial eclipse	1919/05/29	11.31.05.0	$60.1°$	$047.8°$
Maximum eclipse	1919/05/29	13.07.10.1	$70.6°$	$358.2°$
End of partial eclipse	1919/05/29	14.44.19.3	$58.9°$	$310.5°$

Deviation of a starlight closest to the Sun (App.Lat Corr for $70.6°$)
Astronomical refraction   DIP/Terrestrial refraction
- 0,35 minutes = - 21 sec.arc   - 2,8 minutes = - 168 sec.arc

Figure 5.2: Eclipse data of eclipsewise.com [17]

Moreover, if calculated taking into account the deviation caused by terrestrial refraction which is value depend on elevation of the place of observation, and always greater than astronomical refraction, as shown in Figure 5.2. deviation is about 189 sec.arc, it's more than 10 times greater than Einstein's prediction.

In Figure 5.2. the terrestrial refraction/DIP value is calculated with the minimum height of eye 2, 5 meters.

1919 Eclipse - Brazil				
Lat : 11.5239° S				
Long: 56.0742° W				

Event	Date	Time (UT)	Alt	Azi
Start of partial eclipse	1919/05/29	10.34.24.1	07.8°	1°
Maximum eclipse	1919/05/29	11.37.42.1	21.7°	061.1°
End of total eclipse	1919/05/29	11.39.38.9	22.1°	060.9°

Deviation of a starlight closest to the Sun (App.Lat Corr for 21.7°)
Astronomical refraction   DIP/Terrestrial refraction
- 2,55 minutes = - 153 sec.arc   - 2,8 minutes = - 168 sec.arc

Figure 5.3: Eclipse data of eclipsewise.com

In Figure 5.3. we find astronomical data of 1919 eclipse in Brazil:

1. Altitude of the Sun at maximum eclipse is $21.7°$

2. Astronomical refraction = - 153 sec.arc

   Terrestrial refraction = - 168 sec.arc

3. Deviation = - 321 sec.arc, it's more than 100 times greater than Einstein's prediction.

Astronomical refraction in Brazil is about 153 sec.arc; it's more than 7 times greater than the value in West-Africa. This proves the invalidity of Einstein's prediction of 1.75 sec.arc without mentioning the altitude of the Sun as the object of observation.

Astronomical data of 1922 solar eclipse in Australia

1922 Eclipse - Australia
Lat : 30.1451° S
Long: 146.25° E

Event	Date	Time (UT)	Alt	Azi
Start of partial eclipse	1922/09/21	05.00.12.4	38.6°	299.3°
Maximum eclipse	1922/09/21	06.10.00.3	24.7°	286.8°
End of total eclipse	1922/09/21	07.12.50.4	11.5°	277.9°

Deviation of a starlight closest to the Sun (App.Lat Corr for 24.7°)
Astronomical refraction   DIP/Terrestrial refraction
- 2,05 minutes = -123 sec.arc   - 2,8 minutes = -168 sec.arc.

Figure 5.4: Eclipse data of eclipsewise.com

Astronomical data of eclipse in Australia, on Sebtember 21, 1922:

1. Altitude of the Sun at maximum eclipse is 24.7^0

2. Astronomical refraction = - 123 sec.arc

   Terrestrial refraction = - 168 sec.arc

3. Deviation = - 291 sec.arc, it's more than 100 times greater than Einstein's prediction.

Astronomical data as shown in Figure 5.4. that prove Einstein's prediction really doesn't work.

Astronomical data of 1929 solar eclipse in Sumatra

1929 Eclipse - Sumatra
Lat : 0.7031° N
Long: 94.2188° E

Event	Date	Time (UT)	Alt	Azi
Start of partial eclipse	1929/05/09	04.36.25.0	65.6°	043.1°
Maximum eclipse	1929/05/09	06.07.00.4	71.6°	343.2°
End of total eclipse	1929/05/09	07.40.21.2	56.6°	303.2°

Deviation of a starlight closest to the Sun (App.Lat Corr for 71.6°)
Astronomical refraction   DIP/Terrestrial refraction
- 0,35 minutes = - 21 sec.arc.   - 2,8 minutes = - 168 sec.arc.

Figure 5.5: Eclipse data of eclipsewise.com

Astronomical data of 1929 solar eclipse in Sumatra, on May 9:

1. Altitude of the Sun at maximum eclipse is $71.6°$

2. Astronomical refraction = - 21 sec.arc

   Terrestrial refraction = - 168 sec.arc

3. Deviation = - 189 sec.arc, it's more than 100 times greater than Einstein's prediction.

With all due respect I must say again: "Einstein's prediction really doesn't work."

Astronomical data of 1952 solar eclipse in Sudan (Africa)

1952 Eclipse - Sudan (Africa)				
Lat : 14.5064^0 N				
Long: 31.5571^0 E				
Event	Date	Time (UT)	Alt	Azi
Start of partial eclipse	1952/02/25	07.40.43.3	46.5^0	121.3^0
Maximum eclipse	1952/02/25	09.06.43.0	61.8^0	147.1^0
End of total eclipse	1952/02/25	10.34.24.1	65.2^0	196.2^0

Deviation of a starlight closest to the Sun (App.Lat Corr for 61.8^0)

Astronomical refraction	DIP/Terrestrial refraction
- 0,55 minutes = - 33 sec.arc.	- 2,8 minutes = - 168 sec.arc.

Figure 5.6: Eclipse data of eclipsewise.com

Astronomical data of 1952 eclipse in Sudan (Africa), on February 25:

1. Altitude of the Sun at maximum eclipse is 61.8^0

2. Astronomical refraction = - 33 sec.arc

   Terrestrial refraction = - 168 sec.arc

3. Deviation = - 201 sec.arc, it's more than 100 times greater than Einstein's prediction.

With all due respect I must say: 'Einstein's prediction does not work again."

From the astronomical data of 1952 solar eclipse as an observer in Sudan may be can see the North Star of Polaris during the maximum solar eclipse. Altitude of Polaris is the same with latitude of Sudan; it about 14.5 degrees.

STAR CORRECTIONS

DIP-Height of Eye above the Sea in meters		Astronomical Refraction	
1.83	- 2,4	14.00	- 4,1
1.99	- 2,5	15.1	- 3,6
2.15	- 2,6	16.48	- 3,2
2.32	- 2,7	17.19	- 3,1
2.50	- 2,8	17.51	- 3,0
		18.26	- 2,9

Figure 5.7: Star Corrections Table

Figure 5.7. shows the value of astronomical refraction for apparent altitude of Polaris: 14.5 degrees is about -3, 7 minutes or -222 sec. arc. This value is greater than astronomical refraction of the star that being closest to the Sun (- 33 sec.arc) as seen in Figure 5.6. It means spacetime is false

Astronomical data of 1973 solar eclipse in Mauritania (Africa)

1973 Eclipse - Mauritania (Africa) Lat : 17.644° N Long: 3.8672° W				
Event	Date	Time (UT)	Alt	Azi
Start of partial eclipse	1973/06/30	10.01.39	64.5°	073.0°
Maximum eclipse	1973/06/30	11.33.51.4	83.5°	030.5°
End of total eclipse	1973/06/30	13.04.57.1	71.2°	290°

Deviation of a starlight closest to the Sun (App.Lat Corr for 83.5°)
Astronomical refraction   DIP/Terrestrial refraction
- 0,15 minutes = - 9 sec.arc.   - 2,8 minutes = -168 sec.arc.

Figure 5.8: Eclipse data of eclipsewise.com.

Astronomical data of 1973 eclipse in Mauritania (Africa), on June 30:

1. Altitude of the Sun at maximum eclipse is $83.5°$

2. Astronomical refraction =   - 9 sec.arc

   Terrestrial refraction   = - 168 sec.arc

3. Deviation = - 177 sec.arc, it's more than 100 times greater than Einstein's prediction.

In the same way as 1952 eclipse in Sudan, from solar elipse of 1973 in Mauritania we can prove spacetime is false.

The important things to be noted, if spacetime is correct, then the deviation of light track of star that further away from the Sun should be smaller than deviation of light track of the star being closest to the Sun.

But the astronomical data of 1973 eclipse shows:

Altitude of Polaris star is about 17.64 degrees;

1. See table in Figure 5.7. astronomical refraction is about -2,95 minutes or -177 sec.arc.
2. DIP/Terrestrial Refraction: - 2,8 sec.arc or - 168 sec.arc.
3. Deviation : - 345 seconds of arc. It's more than 100 times greater than deviation of the star that being closest to the Sun as seen in Figure 5.8; on the other words, it's not smaller than light track of the star being closest to the Sun.

So, astronomical data of 1973 eclipse in Mauritania (West-Africa) shows crearly that Einstein's prediction does not work, and spacetime is false.

If hypothesis is not valid and spacetime is false, what's then?

Einstein himself had stated clearly and uniquely:
"The chief attraction of the theory lies in its logical completeness. If a single one of the conclusions drawn from it proves wrong, it must be given up; to modify it without destroying the whole structure seems to be impossible."

Einstein never received a Nobel Prize for Relativity

"It doesn't matter Karl Schwarzschild discovered the solution to Einstein's equations which describes black hole, or not; and it doesn't matter Arthur Eddington and Einstein himself didn't believe in black holes, or not; if astronomical data shows very clearly nothing about warping of spacetime, it mean black hole can't happen."

Stephen Hawking's writings and statements are often surprising. As a world-leading physicist, not infrequently his writings and statements make a bit of a stir in the science world. If not he who says; it must be considered as a joke. But because he said that; then it is considered serious and gets more attention. For example, the writings and statements of Stephen Hawking in Nature in the early of 2014: 'There are No black holes'. A number of physicists and astrophysicists reacted strongly against his idea.

Why Hawking is wrong about Black Holes. A recent paper by Stephen Hawking has created quite a stir, even leading Nature News to declare there are no black holes. As I wrote in an earlier post, that isn't quite what Hawking claimed. But it is now clear that Hawking's claim about black holes is wrong because the paradox he tries to address isn't a paradox after all.

So there's no paradox. Black holes can radiate in a way that agrees with thermodynamics, and the region near the event horizon doesn't have a firewall, just as general relativity requires. So Hawking's proposal is a solution to a problem that doesn't exist.What I've presented here is a very rough overview of the situation. I've glossed over some of the more subtle aspects. For a more detailed (and remarkably clear) overview check out Ethan Seigel's post on his blog Starts With a Bang! Also check out the post on Sabine Hossenfelder's blog, Back Reaction, where she talks about the issue herself.(Prof.Brian Koberlein) [18]

Another post in PbsOrg:

What Hawking meant when he said 'there are no black holes'. In a nutshell, Hawking seems to be saying this: instead of an event horizon, there is something else he calls an "apparent horizon." In this apparent horizon, matter and energy is temporarily suspended, but then released. If this is true, it changes black holes as we know them.

"The absence of event horizons means that there are no black holes—in the sense of regimes from which light can't escape to infinity," Hawking wrote in his paper. According to his proposal, black holes do trap information for a long time, but that information can, eventually, escape, Polchinski said. He added that Hawking's proposal remains untested. [19]

Every human has limit. Stephen Hawking has limit. His writing might be right, or might be wrong. It seems he often turned his ideas, for example as it is seen in his famous book 'A Brief History of Time'.

About test of general relativity in the year 1919 Stephen Hawking wrote 'Their measurement had been sheer luck, or a case of knowing the result they wanted to get, not an uncommon occurrence in science'.

Hawking reports the widespread view that the errors in the data were as big as the effect they were meant to probe. Some go further, saying that Eddington deliberately excluded data that didn't agree with Einstein's prediction.(nature)

This is in accordance with his writing in the previous paragraph; discussing general theory of relativity and quantum mechanics Stephen Hawking wrote 'Unfortunately, however, these two theories are known to be inconsistent with each other—they cannot both be correct'.

It shows his view on general theory of relativity. Surprisingly, he developed his idea of black holes based on general theory of relativity.

Logical Fallacies Of Special And General Theory Of Relativity

More surprising, in his book A Brief History of Time he wrote:

"General relativity predicts that heavy objects that are moving will cause the emission of gravitational waves, ripples in the curvature of space that travel at the speed of light. These are similar to light waves, which are ripples of the electromagnetic field, but they are much harder to detect.

The effect of the energy loss will be to change the orbit of the earth so that gradually it gets nearer and nearer to the sun, eventually collides with it, and settles down to a stationary state. The rate of energy loss in the case of the earth and the sun is very low—about enough to run a small electric heater. This means it will take about a thousand million million million million years for the earth to run into the sun, so there's no immediate cause for worry! The change in the orbit of the earth is too slow to be observed, but this same effect has been observed over the past few years occurring in the system called PSR 1913 + 1 6 (PSR stands for "pulsar," a special type of neutron star that emits regular pulses of radio waves). This system contains two neutron stars orbiting each other, and the energy they are losing by the emission of gravitational waves is causing them to spiral in toward each other. This confirmation of general relativity won J. H. Taylor and R. A. Hulse the Nobel Prize in 1993. It will take about three hundred million years for them to collide. " (Stephen Hawking) [20]

A confirmation of the general relativity won the Nobel Prize! It was said by Stephen Hawking. That was what had been expected since 1921; Einstein received a Nobel prize in 1921 for photoelectric effect, not for relativity. At that time the Nobel Committe wrote:

"without taking into account the value that will be accorded your relativity and gravitation theories after these are confirmed in the future"

What Stephen Hawking wrote above was sort of hopeful, or may be a joke. In facts, the Nobel Prize in Physics was awarded to Taylor and Hulse in 1993 for the discovery of a new type of pulsar; without taking into account the value of general relativity.

1993 Nobel Prize in Physics "for the discovery of a new type of pulsar, a discovery that has opened up new possibilities for the study of gravitation" [21]

Why do they still expect the Nobel Prize for Einstein's gravity? It's very obvious the theory was totally wrong. All methods of experiments in any way: i.e. VLBI, Gravity Probe B, LIGO's Twin Detector, Event Horizon Telescope; will be found their mistakes in the experiments. General theory of relativity has been wrong since the beginning and the past can not be erasing and can not be updated.

Open Letter to the Nobel Committee for Physics 2016, W.W. Engelhardt, JET, Max-Planck-Institut für Plasmaphysik. Abstract: The Nobel Committee is informed that according to Professor Karsten Danzmann (Albert Einstein Institut) the LIGO detectors are not calibrated as expected from the statement in the discovery paper: "The detector output is calibrated in strain by measuring its response to test mass motion induced by photon pressure from a modulated calibration laser beam [63]". The claim that gravitational waves have been detected is not substantiated experimentally, since direct calibration data, namely mirror displacement as a function of laser power moving the mirrors, are not published. [22]

"General relativity is Einstein's law of gravity, his explanation of that fundamental force which holds us to the surface of the Earth. Gravity, Einstein asserted, is caused by a warping of space and time—or, in a language we physicists prefer, by a warping of spacetime. The Earth's matter produces the warpage, and that warpage in turn is manifest by gravity's inward tug, toward the Earth's center." (Kip Thorne: **Warping spacetime**) [23]

It doesn't matter Karl Schwarzschild discovered the solution to Einstein's equations which describes black hole, or not; if astronomical data shows very clearly nothing about warping of spacetime, it mean black hole can't happen.

Why didn't they know nonsense?

*T*he first calculation of the deflection of light by mass was published by the German astronomer Johann Georg von Soldner in 1801. Soldner showed that rays from a distant star skimming the Sun's surface would be deflected through an angle of about 0.9 seconds of arc, or one quarter of a thousandth of a degree. This angle corresponds to the apparent diameter of a compact disc (CD) viewed from a distance of about 30 kilometers (nearly 20 miles). Soldner's calculations were based on Newton's laws of motion and gravitation, and the assumption that light behaves like very fast moving particles. As far as we know, neither Soldner nor later astronomers attempted to verify this prediction, and for good reason: Such an attempt would have been far beyond the capability of early 19th century astronomical instruments.

Light deflection in general relativity. Over a century later, in the early 20th century, Einstein developed his theory of general relativity. Einstein calculated that the deflection predicted by his theory would be twice the Newtonian value. The following image shows the deflection of light rays that pass close to a spherical mass. To make the effect visible, this mass was chosen to have the same value as the Sun's but to have a diameter five thousand times smaller (i.e., a density 125 billion times larger) than the Sun's. [24]

How, exactly, did Newton fail? We know that Einstein's general relativity is superior to Newton's gravity, but where did Newton go wrong?

These were small numbers, but a joint expedition by Arthur Eddington and Andrew Crommelin during the 1919 solar eclipse, were able to measure to the necessary accuracy. The deflection they came up with was 1.61" ± 0.30", which agreed (within the errors) with Einstein's predictions, and not with Newton's. Newtonian gravity was busted.

And that's the story of not only Newton's gravity being superseded, but in what way(s) Newton's theory came up short. There have been many other victories for general relativity since (and, honestly, no failures as of yet), but in all the cases where Newton's and Einstein's theories differ, it's Einstein—with stronger gravitational effects close to massive bodies—who emerges victorious. Science marches forward, but sometimes each new step takes a very long time! (Ethan Siegel) [25 }

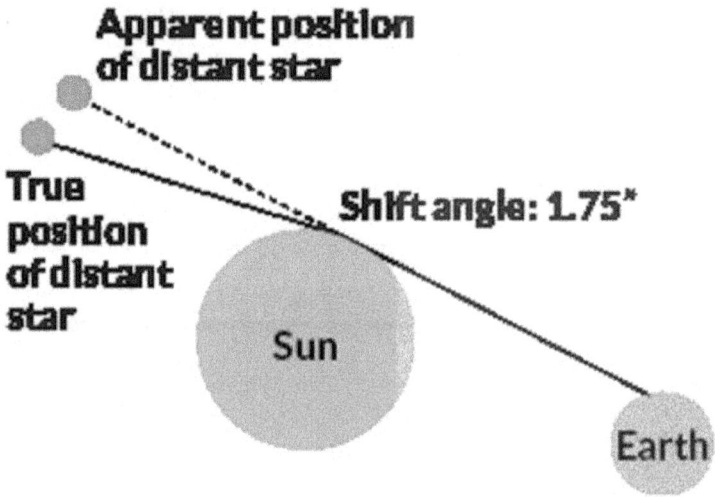

Figure 7.1: Bending light

It is interesting to note, first, on the einstein-online that says 'Einstein calculated that the deflection predicted by his theory would be twice the Newtonian value.'. Second, on the sciencenews.org that explains the bending of light as seen from Earth; shift angle 1.75 seconds of arc.[26]

First, Newton himself never predicts deflection of light by the Sun; to be more precisely, absence of evidence about it. Newton never mentioned the bending of light is about 0.9 seconds of arc.

A paper of Domingos S.L.Soares, 'Newtonian gravitational deflection of light revisited', submitted on 3 Aug 2005; describes the angle of deflection of a light ray by the gravitational field of the Sun, at grazing incidence, is calculated by strict and straightforward classical Newtonian means using the corpuscular model of light. The calculation is presented in the historical and scientific contexts of Newton's {\it Opticks} and of modern views of the problem. [27]

Domingos S.L.Soares drew a conlusion, there is still the question of why Newton did not discuss the possibility of light ray deflection by a massive heavenly body. Of course, he was well acquainted with the relevant astronomical observations.Solar eclipses were certainly of his knowledge and could certainly motivate digressions on the gravitational bending of light.

Second, general relativity also predicts that gravity should bend light, but for very different reasons. In general relativity, a massive object distorts spacetime itself, and light simply takes the straightest path. You have to work through the numbers, but if you do, you discover that this means light bends twice as much in general relativity as in Newtonian gravity.[28]

According to Einstein, the star light visible around the sun would be bent inwards, toward the sun at the time when passing through the gravity field of the sun. Einstein calculated the level of their deviation and predicted that for the stars observed being the closest to the Sun, their deviation was about 1.75 seconds of arc-See Figure 7.1.

For more than 100 years all physicists and astrophysicists are very familiar with the illustration in Figure 1 above; but, did they realize that the above illustration has no meaning whatsoever? *Has no meaning is the same with nonsense.* The above illustration shows exactly how Einstein has been a failure to understand of astronomy.

What is the reason?

This prediction is not meaningful in scientifically of astronomy when it is not explained the altitude of the Sun.

The important **things to be noted**, the amount of 1.75 seconds of arc without mentioning the altitude of the Sun as the object of observation. This is a fatal mistake; or something like a joke; **because the deviation of starlight will always vary depending on the altitude of the object of observation.**

Easy way to know prediction of general relativity really doesn't work.

Albert Einstein may be a genius, or may be Einstein's genius is overrated. But he is a human; every human has limit. It does not matter someone has a college degree or not, as a human can make a mistake in his life. Sometimes those with a college degree like to appear as a fool. For example, if a person who is known as an expert in physics says "I'm certain, Einstein has always been right". Of course, his statement made him look stupid.

We'll easily be able to understand Einstein's theory of gravity was totally wrong; even if our education advanced only through high school. Why so, and what's the reason? That's because Einstein's mistakes are also at the elementary level: *he had no idea on the basic of astronomy.*

1. Prediction of General Relativity

If Einstein's theory of gravity or general theory of relativity was correct, then the light from stars that passed closest to the sun would show the greatest degree of bending, and the stars whose light tracks are very far from the sun have their lights not being bent or deflected. The stars whose lights are not deflected means that there is no difference between the apparent position and the actual position of the stars. It means Einstein's theory of gravity predicts the condition of negligible difference between actual position and apparent position of star in the sky.

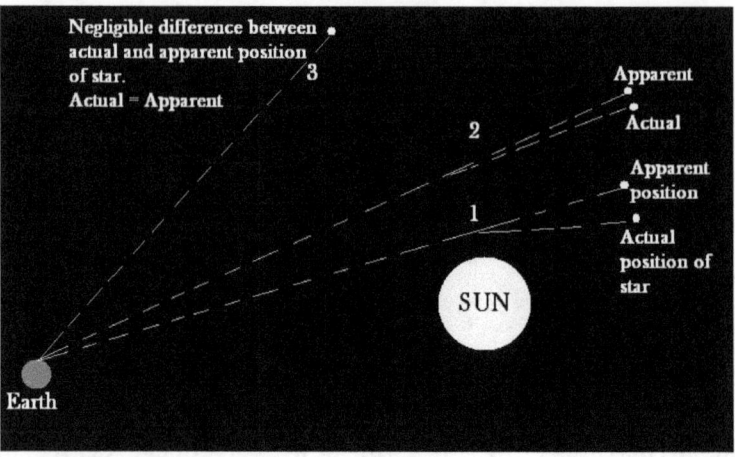

Figure 8.1: General relativity predicts negligible difference between actual and apparent position of star.

Let's look at Figure 8.1 above that deviation of star 1 must be greater than deviation of star 2. And then, star 3 its position is very far from 'warping of spacetime' around the Sun; according to Einstein the light track are not bent. On other word, negligible difference between actual position and apparent position of star.

If being consistent with this theory, it means that all stars visible at night time are at the appearance of the stars on their actual position, because the said stars do not pass through 'the curvature of spacetime' around the Sun. This is certainly incorrect if it is seen from the astronomical scientific point of view. The stars in the sky at night time and seen by the observer, all are stars on apparent position, not on their actual position.

In this case, prediction of general relativity really doesn't work.

2.Einstein's gravity: nothing about force

According to Einstein, gravity is nothing about force; but a curvature spacetime. Einstein had no idea that the fabric of spacetime which curves around the mass of the Sun, Earth and planets is not the empty vacuum but the atmospheric medium.

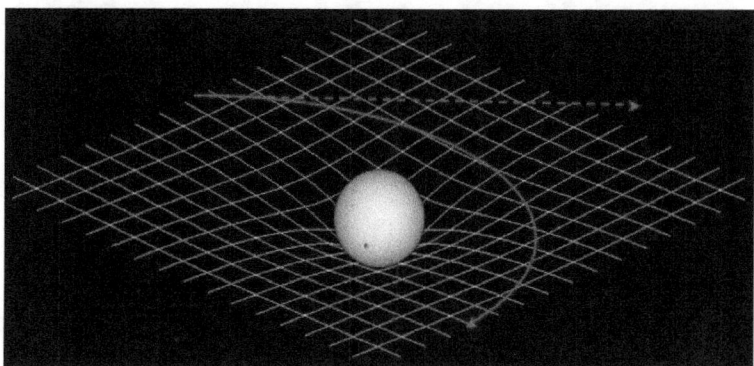

Figure 8.2: A massive object bends spacetime, which deflects a light ray (red) towards the object

If a massive object bends spacetime as show in Figure 8.2 above; suppose the massive object is the Earth, here many of questions arise: What's happen with the Earth's atmospheric medium? Was a curvature spacetime equal to atmospheric medium? If not, why did satellite Gravity Probe B orbits in Earth's atmosphere in order to prove the existence of spacetime? Read more on: Gravity Probe B: Mission Impossible?

Atmospheric medium of the Earth are troposphere (0–16 Km), stratosphere (16–50 Km), mesosphere (50–80 Km), thermosphere or ionosphere (80–640 Km), and exosphere (640–9600 Km), see in Figure 8.3 below.

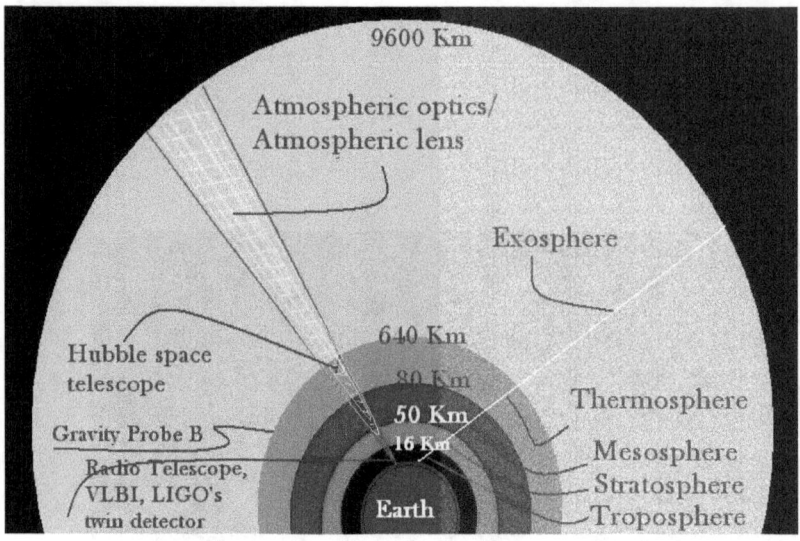

Figure 8.3: Atmospheric Optics

The Earth's atmospheric medium limits the observations of celestial bodies performed by astronomers and astrophysicists. Inevitably, likes or dislikes, their observations -by Hubble space telescope, Radio Telescope/VLBI, LIGO's twin detector, Event Horizon Telescope-are influenced by the atmospheric medium, both the atmospheric medium of the Earth and the atmospheric medium of the object that being observed. Inevitably, likes or dislikes.

It doesn't matter Karl Schwarzschild discovered the solution to Einstein's equations which describes black hole, or not. It doesn't matter in his 1939 paper Einstein credits his renewed concern about the Schwarzschild radius to discussions with Princeton cosmologist Harold P. Robertson and with his assistant Peter G. Bergmann, or not. It doesn't matter in 1994 Hubble Space Telescope provides best evidence to date of super massive black holes that lurk in the center of some galaxies, or not. If astronomial data shows very clearly nothing about warping of spacetime, it means black hole can't happen; and it means 'Einstein's gravity: nothing about force' is false.

I think, at the time of Einstein era is not known much about atmospheric optics. In facts, in special and general theory of relativity

Einstein has ignored the effects of atmospheric medium; to be more precisely Einstein had no ideas in the basic of astronomy.

Equivalence Principle.

Equivalence Principle stating that at any point of space-time the effects of a gravitational field cannot be experimentally distinguished from those due to an accelerated frame of reference. Einstein online explains on this principle in the following.

Imagine you are in an elevator or, more precisely, in what looks like an elevator cabin from the inside, and that you are isolated from the outside world. If you pick up an object and let it drop, it falls down to the floor, in exactly the way you would expecte given your experiences here on Earth. Does that mean the elevator is indeed situated in a gravitational field like that of the Earth-as shown in the following illustration?

It ain't necessarily so. Theoretically, you could be in deep space, far away from all significant mass concentrations and their gravitational influence. The room you are in could be a cabin aboard a rocket—as long as the rocket engines work at exactly the right rate to accelerate the rocket at 9.81 meters per square second. This is sketched in the following illustration-Figure 9.1.

Logical Fallacies Of Special And General Theory Of Relativity

Tom as an observer standing on the cabin floor

Figure 9.1: Inside a rocket, Tom cannot decide whether or not in a gravitational field or not.

Thus, inside an elevator, we cannot decide whether or not we are in a gravitational field or not. Whether or not objects accelerate towards the floor is a matter of reference frame: Even in a gravity-free region of space, objects fall towards the floor if the room we are in is being accelerated. Conversely, even in a gravitational field, objects drift weightlessly through space, as long as the elevator is in free fall.

Einstein became convinced that this inability to distinguish a region with a gravitational field from one without was not just restricted to observations of falling bodies. He postulated that it holds true for any physical measurements at all: No experiment, no clever exploitation of the laws of physics, he claimed, can tell us whether we are in free space or in a gravitational field. This statement is called the equivalence principle.

Einstein's 'happiest thought'

Einstein's happiest thought led to the Equivalence Principle 108 years ago.

While pondering the nature of gravity, Einstein had a sudden revelation, which he would later dub his "happiest thought."

Imagine a workman standing on the roof of a house and losing his footing. As he plummeted in free fall, everything within his grasp (a toolbox, for example) would plunge with him. Therefore, from his local perspective gravity wouldn't seem to exist.

The Equivalence Principle allows us to make calculations involving gravity by considering instead reference frames undergoing acceleration. By applying the Equivalence Principle, the answer equally applies to a reference frame with the same gravitational acceleration. Let's see how ...(phys.ufl.edu-lectures)

For more than 100 years there are many physicists and astrophysicists believe in equivalence principle without a doubt; that is something astonishing facts and surprising. Equivalence principle explains the condition of object falling looks very well, but actually are not well enough. Because the effects of gravity have a quantity; the effects are not merely a falling object on the surface of the Earth.

All objects on Earth are affected by the force of Earth's gravity. The strength of Earth's gravity varies depending on location. The strength of Earth's gravity on the poles of the Earth and on the equator region is not the same. Experience shows that the force of Earth's gravitational force in the sea area near the equator is not the same. This experience is known in the case of increasing or decreasing the speed of the ship; and it is not caused by wind and current factors. This is all unknown in the time when Einstein's 'happiest thought'!

Variation in gravity and apparent gravity

Effective gravity on the Earth's surface varies by around 0.7%, from 9.7639 m/s2 on the Nevado Huascarán mountain in Peru to 9.8337 m/s2 at the surface of the Arctic Ocean. In large cities, it ranges from 9.766 in Kuala Lumpur, Mexico City, and Singapore to

9.825 in Oslo and Helsinki.(Wikipedia)

Therefore, Einstein's happiest thought was the "ancient time thinking". Moreover, if associated with a new discovery of Gravity wave (not ripples in spacetime/gravitational waves); equivalence principle of Einstein is useless.

Einstein's explanation on his original idea of the equivalence principle, as we can read in the book 'The Universe and Dr.Einstein', by Lincoln Barnett, are incomplete. It is very clear there has been a mistake in logic:Logical fallacies of Einstein's theory.. *That is fallacy of composition, assuming that something true of part of a whole must also be true of the whole.*

Einstein's general relativity (Einstein's Field Equation of Gravitation) nothing more than the model which contain ideas about equivalence principle in spacetime that do away or eliminate the existing of atmospheric medium around the massive bodies like the Sun, Stars, and Planets.

"In every case to make enumerations so complete, and reviews so general, that I might be assured that nothing was omitted." (Rene Descartes).

What happens if Gravity Probe B had succeeded test General Relativity?

On 4 May 2011, NASA announced the long-awaited results of Gravity Probe (GP-B), and a month later the results appeared in Phys. After more than 47 years and 750 million dollars, GP-B had succeeded in measuring the general relativistic geodetic and frame-dragging effects on orbiting gyroscopes. In this focus issue, CQG publishes a set of refereed papers that provide the complete details of the experiment, from design of the spacecraft to the final data analysis, thus bringing to a close an extraordinary chapter in experimental gravitation.

Figure 10.1: The Gravity Probe B experiment was carried out in collaboration between Stanford University, NASA, Lockheed Martin and KACST.(einstein.stanford.edu)

As already known, the satellite Gravity Probe B orbiting at the altitude 400 miles (642 km) above the Earth or in the exosphere; they measuring geodetic effect and frame dragging in the atmospheric medium of Earth. If Gravity Probe B had succeeded in measuring, it means Gravity Probe B had proven that spacetime is the same with atmospheric medium.

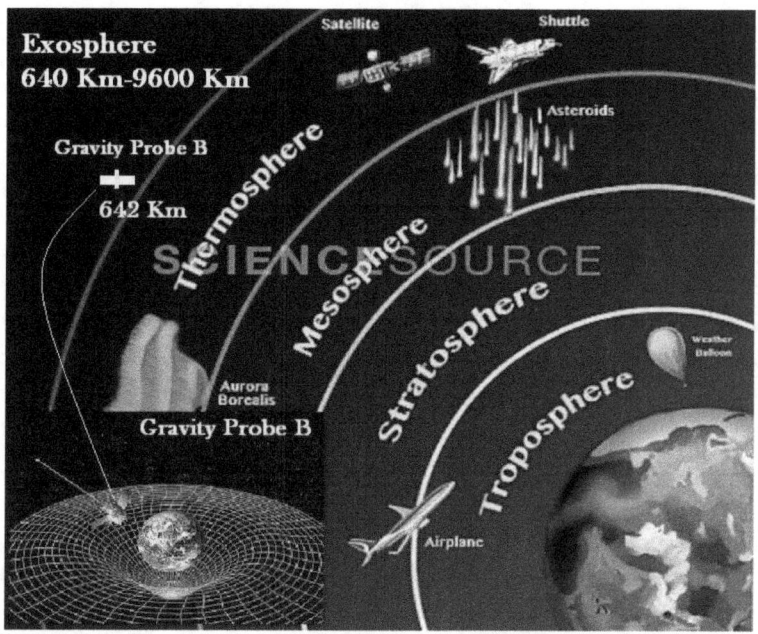

Figure 10.2: Spacetime = Atmospheric medium (science source.com)

Professor Brian Koberlein wrote in Medium: " In 2011 a spacecraft known as Gravity Probe B successfully observed this effect due to the Earth."(The Strangest Theory We Know Is True. After 99 years, Einstein's greatest scientific achievement is undefeated).

If that is true, then the consequence is the acknowledgment that spacetime = atmospheric medium. But, that is become a puzzle or deviate from each other. On one side, it says that test of general relativity succeeded. On the other hand, it proven that spacetime is the same with atmospheric medium; and this means general relativity is wrong.

If it says test of general relativity succeeded, this is very interesting, because it will cause many other logical consequences.

What are the other consequences?

First, if there are no spacetime but atmospheric medium, then there are no black holes. In this case, Stephen Hawking was correct as he states in Nature 'There are no black holes'

Most physicists foolhardy enough to write a paper claiming that "there are no black holes"—at least not in the sense we usually imagine—would probably be dismissed as cranks. But when the call to redefine these cosmic crunchers comes from Stephen Hawking, it's worth taking notice. In a paper posted online, the physicist, based at the University of Cambridge, UK, and one of the creators of modern black-hole theory, does away with the notion of an event horizon, the invisible boundary thought to shroud every black hole, beyond which nothing, not even light, can escape.

"There is no escape from a black hole in classical theory," Hawking told Nature. Quantum theory, however, "enables energy and information to escape from a black hole". A full explanation of the process, the physicist admits, would require a theory that successfully merges gravity with the other fundamental forces of nature. But that is a goal that has eluded physicists for nearly a century. "The correct treatment," Hawking says, "remain a mystery"

Stephen Hawking: 'Black Holes Ain't As Black As They Were Painted'. This has been an outstanding problem in theoretical physics for the last 40 years," Hawking told a much smaller audience of top physicists at the conference, organized by University of North Carolina physicist Laura Mersini-Houghton and dedicated to tackling this paradox. As of yet, Hawking said, "no satisfactory resolution has been advanced."

At the core of the information paradox is the tension between two theories of nature, which are "perhaps the two most successful theories in the history of science," says Dan Hooper, a theoretical astrophysicist at Fermilab.

On one hand is general relativity, which says that anything that enters a black hole is lost in the immense power of its gravity and that a black hole eventually destroys any information—location, velocity and orientation, for example—about what it has swallowed up, including the star that birthed the black hole. On the other hand stands the theory of quantum mechanics, which says that information can never be lost.(newsweek.com).

Figure 10.3: This is not a photo of a black hole. This is an artist's depiction of how matter might be ripped and radiate energy as it orbits the black hole's high-gravity center (These Are Not Photos of Black Holes, gizmodo.com).

Black Holes. Black holes may be heard but not seen. As the death state of a star, black holes emit no light. They are dark against the dark sky. But like mallets on a drum, black holes can ring out a song on space itself in the form of gravitational waves. Monumental Earth-based detectors such as LIGO and the planned space-based mission LISA aim to record the songs from space for the first time thereby

turning up the volume on the soundtrack to the Universe.
There is no sound in empty space. But when the gravitational waves hit the Earth, detectors in the next few years will measure the stretching and shrinking of space and will be able to amplify the result as sound. (Janna Levin's science).

Black holes do NOT exist and the Big Bang Theory is wrong, claims scientist—and she has the maths to prove it. Scientist claims she has mathematical proof black holes cannot exist. Professor Laura Mercini-Houghton said she is still in 'shock' from the find (Black holes do NOT exist and the Big Bang Theory is wrong, claims scientist - and she has the maths to prove it, dailymail).

Second, if there are no spacetime but atmospheric medium, then we need the new physics ..

3. What Exactly is Gravity?

How did our Universe get to be so homogenous, beautiful and special? Gravity is the builder of the balance Universe and all the celestial bodies

How did the structure of Earth's atmosphere get to be so regular? The formation of Earth's atmosphere was caused by Earth's gravity.
Gravity isn't merely the law of attraction, Earth's gravity is the builder of the Earth's atmosphere.

Gravity compresses the Earth's atmosphere, it creates air pressure- the driving force of wind. Without gravity, there would be no atmosphere or air pressure and thus, no wind. It's mean there would be no life at the Earth.

Planet's gravity is the builder of planet's atmosphere.

In the history of gravity we can read on Wikipedia that in the 4th century BC, the Greek philosopher Aristotle believed that there is no effect or motion without a cause. The cause of the downward motion of heavy bodies, such as the element earth, was related to their nature, which caused them to move downward toward the center of the universe, which was their natural place.

Conversely, light bodies such as the element fire, move by their nature upward toward the inner surface of the sphere of the Moon. Thus in Aristotle's system heavy bodies are not attracted to the earth by an external force of gravity, but tend toward the center of the universe because of an inner gravitas or heaviness.

Brahmagupta, the Indian astronomer and mathematician whose work influenced Arab mathematics in the 9th century, held the view that the earth was spherical and that it attracted objects. Al Hamdānī and Al Biruni quote Brahmagupta saying "Disregarding this, we say that the earth on all its sides is the same; all people on the earth stand upright, and all heavy things fall down to the earth by a law of nature, for it is the nature of the earth to attract and to keep things, as it is the nature of water to flow, that of fire to burn, and that of the wind to set in motion.

If a thing wants to go deeper down than the earth, let it try. The earth is the only low thing, and seeds always return to it, in whatever direction you may throw them away, and never rise upwards from the earth.

Isaac Newton

During the 17th century, Galileo found that, counter to Aristotle's teachings, all objects accelerated equally when falling.

In the late 17th century, as a result of Robert Hooke's suggestion that there is a gravitational force which depends on the inverse suare of the distance. Isaac Newton was able to mathematically derive Kepler's three kinematic laws of planetary motion including the eclliptical orbits for the six then known planets and the Moon:

"I deduced that the forces which keep the planets in their orbs must be reciprocally as the squares of their distances from the centres about which they revolve, and thereby compared the force requisite to keep the moon in her orb with the force of gravity at the surface of the earth and found them to answer pretty nearly."
-Isaac Newton, 1666

So Newton's original formula was:

$$\text{Force of gravity} \propto \frac{\text{mass of object 1} \times \text{mass of object 2}}{\text{distance from centers}^2}$$

where the symbol: $\propto$ means "is proportional to"

What is Gravity?

According to Livescience magazine gravity is the force that attracts two bodies toward each other, the force that causes apples to fall toward the ground and the planets to orbit the sun. The more massive an object is, the stronger its gravitational pull.

But by definition gravity is the force that attracts a body towards the centre of the earth, or towards any other physical body having mass (Oxford Dictionary).

Newton's law of universal gravitation states that any two bodies in the universe attract each other with a force that is directly proportional to the product of their masses and inversely proportional to the square of the distance between them.

In this case we see that our understanding of gravity is not complete. We need to think about gravity in the large scale of the universe and in relation with balance of universe. The universe is always in a state of harmony and balance - homogenous and isotropic of the universe—from somewhere on earth look to the sky, all point in the sky always looks the same and uniform. Gravity not only a force that any two bodies in universe attract each other, but gravity is the force due to the effects of well balanced universe.

The universe demands balance. Nature demands balance. All of celestial bodies in orbit, demand balance. Earth's rotation demands balance. Life demands balance

The Law of Compensation is that for every action, there is an equal and opposite reaction. If the pendulum swings one way, it must always swing back the other way. If you will, for a moment, stop; and consider that these swings of rhythm are also evidenced in your life. Courage is preceded by fear. Happiness and sadness oscillate.

The ancients say, "The man who enjoys well can also be subject to great suffering. The man who feels little pain is capable of feeling but little joy."

Mentally, however, it is possible to escape the low end by rising above it. We are able to overcome the swing of the pendulum by making the vibrations higher and rising above the lower vibrations. In essence, we are raising the vibrations of the self above the ordinary plane of consciousness, and then simply "refusing" to allow the pendulum of emotion and mood to swing us back.

Even so, the Law of Compensation is operative. You will probably find that there is no such thing as an overnight success. One generally pays the price for what he wants to attain. The things that one pays a price for are always repaid.

Always remember that everything is subject to the principle of cause and effect. There is a cause for every effect, and vice versa. Regardless of a belief that says that there is no cause and effect, the principle is always operative.(execonn.com)

There are many phenomena we have found how the effects of well balanced universe really works, and it is also evidence to the thought of Aristotle 'there is no effect or motion without a cause':

1.Gravity and Atmosphere

Gravity is what keeps a planet's gaseous atmosphere from spreading out into space away from the planet. If we compare the gravitational pull of each planet in our solar system we would find them to be different. This is because a planet's gravity is related to its mass. Usually the greater a planet's mass, the greater the gravitational pull.

2.Gravity and Tides

The gravitational attraction between the Earth and the moon is strongest on the side of the Earth that happens to be facing the moon, simply because it is closer.

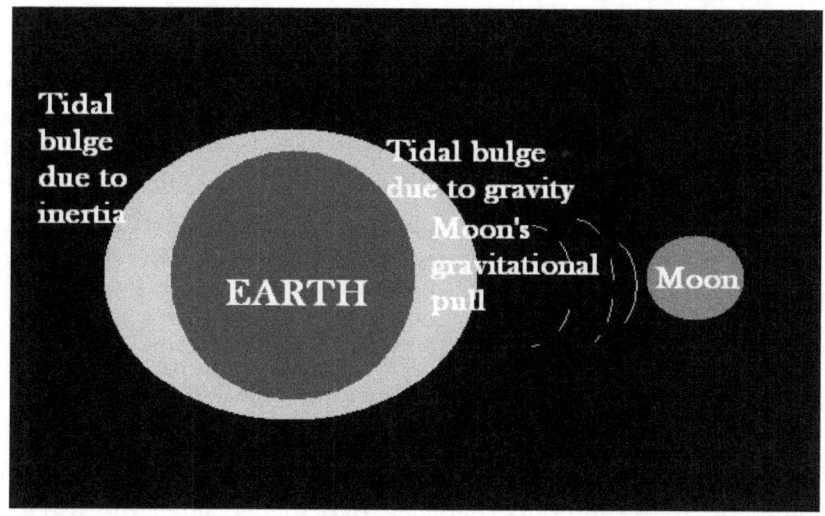

Figure 11.1: Gravity and inertia act in opposition on the Earth's oceans, creating tidal bulges on opposite sites of the planet. On the "near" side of the Earth (the side facing the moon), the gravitational force of the moon pulls the ocean's waters toward it, creating one bulge. On the far side of the Earth, inertia dominates, creating a second bulge. (oceanservice.noaa.gov)

This attraction causes the water on this "near side" of Earth to be pulled toward the moon. As gravitational force acts to draw the water closer to the moon, inertia attempts to keep the water in place. But the gravitational force exceeds it and the water is pulled toward the moon, causing a "bulge" of water on the near side toward the moon.

3. Gravity and Earthquake

The relation between plate tectonics and earthquake evolution is analyzed systematically on the basis of 1998–2010 absolute and relative gravity data from the Crustal Movement Observation Network of China. Most earthquakes originated in the plate boundary or within the fault zone. Tectonic deformation was most intense and exhibited discontinuity within the tectonically active fault zone because of the differential movement; the stress accumulation produced an abrupt gravity change, which was further enhanced by the earthquake. The gravity data from mainland China since 2000 obviously reflected five major earthquakes (Ms > 7), all of which were better reflected than before 2000. Regional gravity anomalies and a gravity gradient change were observed in the area around the epicenter about 2 or 3 years before the earthquake occurred, suggesting that gravity change may be a seismic precursor. Furthermore, in this study, the medium-term predictions of the Ms7.3 Yutian, Ms8.0 Wenchuan, and Ms7.0 Lushan earthquakes are analytically presented and evaluated, especially to estimate location of earthquake. (sciencedirect.com).

4. Gravity Waves, gravity in fluid dinamics

In fluid dynamics, gravity waves are waves generated in a fluid medium or at the interface between two media when the force of gravity or buoyancy tries to restore equilibrium. An example of such an interface is that between the atmosphere and the ocean, which gives rise to wind waves.

When a fluid element is displaced on an interface or internally to a region with a different density, gravity will try to restore it toward equilibrium, resulting in an oscillation about the equilibrium state or wave orbit.

Gravity waves on an air–sea interface of the ocean are called surface gravity waves or surface waves, while gravity waves that are within the body of the water (such as between parts of different densities) are called internal waves. Wind-generated waves on the water surface are examples of gravity waves, as are tsunamis and ocean tides.

Wind-generated gravity waves on the free surface of the Earth's ponds, lakes, seas and oceans have a period of between 0.3 and 30 seconds (3 Hz to 0.03 Hz). Shorter waves are also affected by surface tension and are called gravity–capillary waves and (if hardly influenced by gravity) capillary waves. Alternatively, so-called infragravity waves, which are due to subharmonic nonlinear wave interaction with the wind waves, have periods longer than the accompanying wind-generated waves.

5. Gravity waves can energize tornados

Gravity waves are global events. Much like the ripples on a

massive pond, these large-scale waves can propagate from an atmospheric disturbance over thousands of miles.

These waves are maintained by the gravitational force of Earth pulling down and the buoyancy of the atmosphere pushing up. Until now it has been hard to link atmospheric gravity waves with other atmospheric phenomena, but new research suggests that gravity waves passing over storms can spin up highly dangerous and damaging tornados... Suddenly gravity waves become very important and may help to forecast where and when tornados may strike.

6.Gravity and Tsunami

A tsunami can be generated by any disturbance that displaces a large water mass from its equilibrium position. Submarine landslides, which often occur during a large earthquake, can also create a tsunami. During a submarine landslide, the equilibrium sea-level is altered by sediment moving along the sea-floor. Gravitational forces then propagate the tsunami given the initial perturbation of the sea-level.

7.Gravity and Katabatic wind

A katabatic wind, is the technical name for a drainage wind, a wind that carries high density air from a higher elevation down a slope under the force of gravity. Such winds are sometimes also called fall winds. Katabatic winds can rush down elevated slopes at hurricane speeds, but most are not as intense as that, and many are of the order of 10 knots (18 km/h) or less.

Not all downslope winds are katabatic. For instance, winds such as the föhn, chinook, and bergwind are rain shadow winds where air driven upslope on the windward side of a mountain range drops its moisture and descends leeward drier and warmer. Examples of true katabatic winds include the bora (or bura) in the Adriatic, the Bohemian Wind or Böhmwind in the Ore Mountains, the Santa Ana in southern California, and the oroshi in Japan. Another example is "the Barber", an enhanced katabatic wind that blows over the town of Greymouth in New Zealand when there is a southeast flow over the South Island. It is a wind that is known in the area for its coldness.

8. Earth wobbles on its axis

Since 2003, Greenland has lost on average more than 272 trillion kilograms of ice a year, and that affects the way the Earth wobbles in a manner similar to a figure skater lifting one leg while spinning, said Nasa scientist Eirk Ivins, the study's co-author.

On top of that, West Antarctica loses 124 trillion kgs of ice and East Antarctica gains about 74 trillion kgs of ice yearly, helping tilt the wobble further, Ivins said.

They all combine to pull polar motion toward the east, Adhikari said.
Jianli Chen, a senior research scientist at the University of Texas' Center for Space Research, first attributed the pole shift to climate change in 2013, and he said this new study takes his work a step further.

"There is nothing to worry about," said Chen, who wasn't part of the Nasa study. "It is just another interesting effect of climate change." (theguardian.com)

Where does Gravity come from?

In fact, the Earth received energy in the form of radiation from the Sun. For the Earth to remain in balance the energy coming into and leaving the Earth must equal. Energy for Earth' gravity come from the Sun. But, where does the Sun's energy come from?

The classical scientists such as Aristotle, Rene Desscrates, Sir Isaac Newton and others believed that the light of the stars reaching us on earth crept spreading through a medium the so-called "luminiferous aether" or "aether/ether". However various kinds of experiments had been made, among other was an experiment conducted by the American Scientists Michelson and Morrey in the 19th century, and all of those experiments failed to detect the presence of luminiferous ether, so that the ether is deemed non-existent. There is a possibility that luminiferous ether truly exists, but it cannot be proven.

If the existence of aether is true, then we can figure that energy from the Sun come from aether. We know that the aether can not be detected, and something that can not be detected does not mean it should be considered non-existent. So, we can say that gravity come from aether.

The only real force in this reality is the vibrating of aether. The forces, as well as energy and matter, are only emergent properties of the vibrating of aether. Empty space doesn't really exist, everywhere is completely full of aether.

4.CONCLUSION

There are at least 5 logical fallacies of Einstein's special and general theory of relativity

1. Einstein's Thought Experiment: Fallacy Of Composition.
2. The Equivalence Principle: A false equivalence
3. Spacetime: A fallacy of ambiguity or reification.
4. Einstein's Field Equation: Logical fallacy : argumentum ad lapidem
5. Einstein's proving method: The fallacy of ignoratio elenchi, or irrelevant conclusion.

Lincoln Barnett's book, The Universe and Dr.Einstein, London, 1949, foreword by Albert Einstein himself, tell us about Einstein's idea of Special and General theory of relativity. Page 78 of this book shows the fact that Einstein had no idea on the basic of astronomy. Einstein's hypothesis of general relativity is not valid, therefore, doesn't meet requirement of the principles of scientific method. Beside that, Einstein proposed test via eclipse is not scientifically correct and deeply wrong.

Hypothesis and Einstein proposed test of general relativity are closely related to astronomy, especially celestial navigation. For understanding that hypothesis and the test are not valid, physics training is needed; but more importantly is celestial navigation training. Unfortunately, physicists and astrophysicists are not trained to become experts in the field of celestial navigation. The navigators around the world will be easily to recognize the fatal flaws of these hypotheses and test. Actually, general relativity can not be proven or tested in any way.

REFERENCES

1. Logical Fallacies of Einstein's Theories.

1. **Lincoln Barnett**, The Universe and Dr.Einstein, London, June 1949.

2. **Michael Suede**, What exactly is spacetime?, July 22, 2014.

3. **The speed of light** and the index refraction, http://www.rpi.edu/dept/phys/

4. **Aleksandar Vukelja**, Mathematical Invalidity of the Lorentz Transformation in Relativity Theory

http://www.masstheory.org/lorentz.pdf

5. **Stuart Clark**, Why Einstein never received a Nobel prize for relativity, The Guardian, 8 October 2012.

6. **Dr.Louis Essen**, Relativity-Joke or Swindle?

http://www.ekkehard-friebe.de/Essen-L.htm

7. **F.W. Dyson**, Determination of the deflection of light by the Sun's Gravitational Field, from Observations made at the Total Eclipse of May 29, 1919, April 20, 1919.

https://archive.org/details/philtrans06337895

8. **Tom Van Flandern Articles**

http://www.ldolphin.org/vanFlandern/

9. **Miles Mathis,** The Perihelion Precession of Mercury

http://milesmathis.com/merc.html

10. The Scientific Method

http://www.scientificpsychic.com/workbook/scientific-method.htm

11. **Professor R. C. Gupta**, India, 'Bending of Light Near a Star and Gravitational Red/Blue Shift: Alternative Explanation Based on Refraction of Light'

https://arxiv.org/ftp/physics/papers/0409/0409124.pdf

12. **Mile Mathis**, Gravity Probe and space-time.

http://milesmathis.com/probe.pdf

13. **Proof of** the Invalidity of the Black Hole and Einstein's Field Equations

http://vixra.org/pdf/1212.0060v1.pdf

14. **Wikipedia**: List of fallacies

https://en.wikipedia.org/wiki/List_of_fallacies

15. .**Bowditch**, American Practical Navigator, Volume I - II, Defense Mapping Agency Hydrographic / Topographic Center, 1984.

16. **John P.Budlong**, Sky and Sextant: Practical Celestial Navigation, Second Edition, Van Nostrand Reinhold Company, New York, 1981.

17. **Wikipedia**: Nautical almanac

https://en.wikipedia.org/wiki/Nautical_almanac

18. **NASA** Eclipse Website

https://eclipse.gsfc.nasa.gov/SEgoogle/SEgoogle2001/SE2017Aug21Tgoogle.html

19.**Earthsky** Website.

http://earthsky.org/astronomy-essentials/august-21-2017-solar-eclipse-4-planets-bright-stars

2. Knowing the result they wanted to get

1.**Gravity Probe B** The elativity Mission
https://www.nasa.gov/mission_pages/gpb/

2.**Brian Koberlein**, The Strangest Theory We Know Is True
https://briankoberlein.com/2015/01/12/strangest-theory-know-true/

3.**New measurement** of solar gravitational deflection of radio signals using VLBI
http://www.nature.com/nature/journal/v349/n6312/abs/349768a0.html

4.**Philip Ball**, Arthur Eddington was innocent!
http://www.nature.com/news/2007/070907/full/news070903-20.html

5.**Sabine Hossenfelder**, A wonderful 100th anniversary gift for Einstein
http://backreaction.blogspot.co.id/2015/04/a-wonderful-100th-anniversary-gift-for.html

6.**Gravitational** deflection of light
http://www.einstein-online.info/spotlights/light_deflection

7.**Sextant**
https://en.wikipedia.org/wiki/Sextant

8. **What are Radio Telescopes?**
https://public.nrao.edu/telescopes/radio-telescopes/

9. **Bending Light**
http://www.mathpages.com/rr/s6-03/6-03.htm

10. **Brian Koberlein**, Why Einstein will never be wrong
https://www.universetoday.com/108044/why-einstein-will-never-be-wrong/

11. **Equatorial** Coordinate System
http://astronomy.swin.edu.au/cosmos/E/Equatorial+Coordinate+System

12. **Einstein was WRONG**: 'Spooky' quantum experiment shows that the measurement of a photon affects its location
http://www.dailymail.co.uk/sciencetech/article-3015318/Einstein-WRONG-Spooky-quantum-experiment-shows-measurement-photon-affects-location.html

13. **Tests of general relativity**
https://en.wikipedia.org/wiki/Tests_of_general_relativity

14. **Illuminating relativity**: Experimenting with the stars
http://undsci.berkeley.edu/article/0_0_0/natural_experiments

15. **Einstein's math**
https://divisbyzero.com/2008/10/13/einsteins-math/

16. **David Tong**, What is general relativity?
https://plus.maths.org/content/what-general-relativity

17. **Solar Eclipse**
http://eclipsewise.com/solar/solar.html

18. **Brian Koberlein**, Why Hawking is wrong about Black Holes
https://www.universetoday.com/108870/why-hawking-is-wrong-about-black-holes/

19. **What Hawking** meant when he said 'there are no black holes'
http://www.pbs.org/newshour/updates/hawking-meant-black-holes/

20. **Stephen Hawking**, A Brief History Of Time
https://archive.org/details/ABriefHistoryOfTime

21. **The Nobel Prize** in Physics 1993
http://www.nobelprize.org/nobel_prizes/physics/laureates/1993/

22. **Open Letter** to the Nobel Committee for Physics 2016

https://www.researchgate.net/publication/304581873_Open_Letter_to_the_Nobel_Committee_for_Physics_2016

23. **Kip Thorne**, Warping spacetime
http://www.ws5.com/spacetime/SpaceTime%20Warp.pdf

24. **Gravitational** deflection of light
http://www.einstein-online.info/spotlights/light_deflection

25. **Ethan Siegel**, #106 How Exactly Did Newton Fail?

https://medium.com/starts-with-a-bang/ask-ethan-106-how-exactly-did-newton-fail-6fe008a38045

26. **Tom Siegfried**, Einstein's genius changed science's perception of gravity

https://www.sciencenews.org/article/einsteins-genius-changed-sciences-perception-gravity?mode=magazine&context=1483

27. **Domingos Soares**, Newtonian gravitational deflection of light revisited

https://www.researchgate.net/publication/2173573_Newtonian_gravitational_deflection_of_light_revisited

28.**Classical Tests** of General Relativity
http://www.thephysicsmill.com/2015/11/28/classical-tests-general-relativity/

ABOUT THE AUTHOR

Capt (Ret) Gatot Soedarto, was born in Tuban, East Java, he graduated from The Indonesian Naval Academy. Experienced as lecturer on astronomy and digital engineering at Indonesian Naval Academy, and lecturer on naval strategy at Indonesian Naval Command and Staff College, as well as instructor on celestial navigation and fire fighting.

The books of his works among others are: Computer Engineering (1981), Sun Tzu and Naval Strategy (2012), Lessons of the Falklands War (2013), Other Views of Naval Battles (2014), Eclipse 1919 and the general relativity theory (2014), Albert Einstein Failed In Three Classical Tests (2016).

Follow twitter: @GatotSoedarto

www.ingramcontent.com/pod-product-compliance
Lightning Source LLC
Chambersburg PA
CBHW061438180526
45170CB00004B/1458